my **revision** notes

D0528303

AQA GCSE

FOOD PREPARATION AND NUTRITION

Alexis Rickus
Bev Saunder
Yvonne Mackey

HODDER EDUCATION
AN HACHETTE UK COMPANY

3011780306151 0

Picture credits: p3 © uckyo - Fotolia.com; p5 © Squareplum – Fotolia; p7 © iStockphoto.com; p11 © Alen-D – Fotolia; p14 © CharlieAJA - iStock via Thinkstock/ Getty Images; p15 © BVDC – Fotolia; p17 © Food Standards Agency (OGL); p18 © Radius Images / Alamy Stock Photo; p19 © Chris Pearsall / Alamy Stock Photo; p20 © pioneer111 - iStock via Thinkstock/Getty Images; p21 © Patrizia Tilly – Fotolia; p22 *l* © Clynt Garnham Food & Drink / Alamy Stock Photo, *c* © Isolated Products (John Boud) / Alamy Stock Photo, *r* © studiomode / Alamy Stock Photo; p23 © monticello - 123RF; p25 © Wavebreak Media ltd / Alamy Stock Photo; p28 *l* © monkeybusinessimages - iStock via Thinkstock/Getty Images, *r* © Joe Gough – Shutterstock; p31 © BSIP SA / Alamy Stock Photo; p34 *t* © anandkrish16/123RF, *b* © Stockbyte/ Getty Images Ltd/ Fast Food SD175; p41 © Andrew Callaghan; p42 © Andrew Callaghan; p43 © Andrew Callaghan; p46 © dpullman - iStock – Thinkstock/ Getty Images; p49 © Andrew Callaghan; p51 EYE OF SCIENCE/SCIENCE PHOTO LIBRARY; p52 © BSIP SA / Alamy Stock Photo; p54 *tl* © Pavel Chernobrivets - 123RF, *tr* © Jultud - iStock via Thinkstock/Getty Images, *c* © Yvonne Mackey, *b* © alicjane - iStock via Thinkstock/Getty Images; p55 *t* © jaroslava V – Shutterstock, *c* © Robert Anthony - 123RF, *b* © margouillat - 123RF; p57 © Ingor Normann – Fotolia; p59 © Andrew Callaghan; p61 © Sean Malyon/Corbis; p62 © praisaeng – Fotolia; p63 © Cathy Yeulet - 123RF; p64 © Andrew Callaghan; p67 © gpointstudio - iStock via Thinkstock/Getty Images; p72 © Patti McConville / Alamy Stock Photo; p76 © glegorly - iStock via Thinkstock/Getty Images; p78 © Joe Gough – Fotolia; p79 © DENIO RIGACCI - iStock via Thinkstock/Getty Images; p80 © glsk – Fotolia; p81 © showcake - iStock via Thinkstock/Getty Images; p82 © MediablitzImages – Fotolia; p86 © Peter Erik Forsberg/Food / Alamy Stock Photo; p87 © andystjohn – fotolia; p89 © iconimage – Fotolia; p93 © JGregorySF - iStock via Thinkstock/Getty Images; p96 © sergbob – Fotolia; p100 © circlePS – Shutterstock; p108 © BWFolsom - iStock via Thinkstock/Getty Images.

First published in 2017
by Hodder Eduction,
An Hachette UK Company

Carmelite House
50 Victoria Embankment
London EC4Y 0DZ

© Yvonne Mackey, Alexis Rickus and Bev Saunder 2017

ISBN 978 1 4718 8699 7

First printed 2017
Impression number   5   4   3   2   1
Year   2022   2021   2020   2019   2018   2017

All rights reserved; no part of this publication may be reproduced, stored in a retrieval system, or transmitted, in any form or by any means, electronic, mechanical, photocopying, recording or otherwise without either the prior written permission of Hodder Education or a licence permitting restricted copying in the United Kingdom issued by the Copyright Licensing Agency Ltd, Saffron House, 6–10 Kirby Street, London EC1N 8TS.

Cover photo reproduced by permission of © Getty Images/Thinkstock/iStock/Denira 777

Typeset in Bembo Std Regular 11/13 pts by Aptara Inc.

Printed in Spain

Hachette UK's policy is to use papers that are natural, renewable and recyclable products and made from wood grown in sustainable forests. The logging and manufacturing processes are expected to conform to the environmental regulations of the country of origin.

Get the most from this book

Everyone has to decide their own revision strategy, but it is essential to review your work, learn key facts and test your understanding. These Revision Notes will help you to do that in a planned way, topic by topic. You can check your progress by ticking off each section as you revise.

Tick to track your progress

Use the revision planner on pages iv and v to plan your revision, topic by topic. Tick each box when you have:
- revised and understood a topic
- tested yourself
- practised exam questions and gone online to check your answers and complete the quick quizzes.

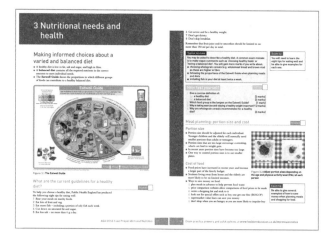

You can also keep track of your revision by ticking off each topic heading in the book. You may find it helpful to add your own notes as you work through each topic.

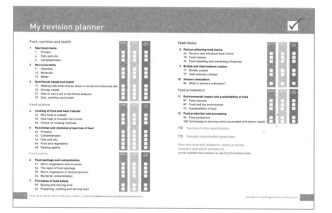

Features to help you succeed

Exam tips

Expert tips are given throughout the book to help you polish your exam technique in order to maximise your chances in the exam.

Typical mistakes

The authors identify the typical mistakes candidates make and explain how you can avoid them.

Now test yourself

These short, knowledge-based questions provide the first step in testing your learning. Answers are at the back of the book.

Key words

Key words from the specification are highlighted in bold throughout the book.

Revision activities

These activities will help you to understand each topic in an interactive way.

Exam practice

Practice exam questions are provided for each topic. Use them to consolidate your revision and practise your exam skills.

Online

Go online to try out the extra quick quizzes at www.hoddereducation.co.uk/myrevisionnotes

My revision planner

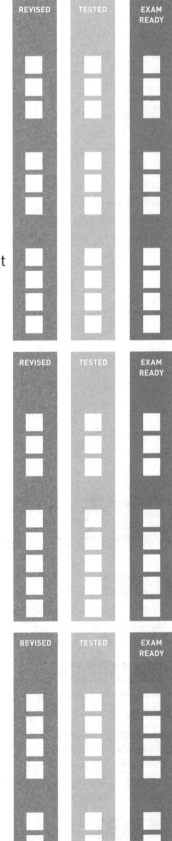

Food choice

Food provenance

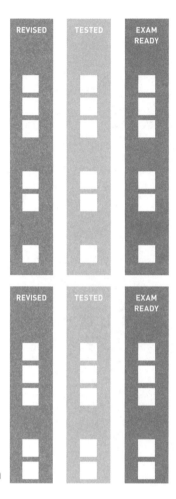

Now test yourself answers, exam practice
answers and quick quizzes at
www.hoddereducation.co.uk/myrevisionnotes

1 Macronutrients

Macronutrients are needed by the body in large amounts. Protein, fat and carbohydrates are macronutrients.

Protein

- Protein is present throughout the human body.
- Protein is a **secondary energy source**.
- One gram of protein provides **4 kcals of energy**.

Functions, sources, deficiency and excess

Table 1.1 **Functions and sources of protein**

Functions in the body	Growth	
	Repair	
	Maintain tissue	
	A secondary source of energy	
Sources of protein	**High biological value (HBV) sources**	**Low biological value (LBV) sources**
	Meat	Cereals (e.g. rice, oats)
	Fish	Wheat
	Eggs	Peas, beans and lentils
	Milk	Nuts and seeds
	Cheese	
	Soya beans	
	Quinoa	
	Mycoprotein (Quorn)	
What happens if we don't get enough protein?	**Children**	**Adults**
	Poor growth	Poor growth
	Thinning hair or hair loss	Fluid under the skin (oedema)
	Catch infections (e.g. colds) easily	Thinning hair or hair lose
	Fluid under their skin (oedema)	Catch infections (e.g. colds) easily
What happens if we get too much protein?	Puts strain on kidneys and liver	
	Increased weight, as extra protein is converted into fat	

> **Exam tip**
>
> There are four functions of protein:
>
> **G**rowth, **E**nergy, **R**epair and **M**aintain = GERM
>
> Note that 'energy' is second in this list. This is because protein is a secondary source of energy; most energy comes from fats and carbohydrates.

The biological value of protein

- Proteins are made up of building blocks called **amino acids**.
- The body can make amino acids but some can only be obtained from food – these are called **essential amino acids**. There are eight essential amino acids needed by adults and children, and at least two more needed just by children because they are growing.
- Foods that contain all the essential amino acids are described as **high biological value (HBV)**. These include soya beans and soya products, and quinoa.
- Foods that lack one or more of the essential amino acids are described as **low biological value (LBV)**.

Protein complementation

- Combining LBV protein foods to form an HBV protein meal is known as **protein complementation**.
- Protein complementation is needed to make sure that vegetarians get all the essential amino acids.
- The essential amino acids lacking in one of the LBV foods will be provided by the other LBV food.
- An example of this is beans on toast. Beans and toast separately are both LBV foods, but together they become HBV protein as, when combined, all of the essential amino acids are present.
- Protein complementation can save money because LBV foods tend to be cheaper than HBV foods.

Revision activity

Match up the pairs below to give some typical examples of protein complementation.

Lentil dhal	Rice
Peas	Chapattis
Baked beans	Pitta bread
Hummus (chickpeas)	Toast

Exam tip

Try to learn **one example** of protein complementation. In 'explain' or 'describe' questions, examples will be credited.

Protein alternatives

- Protein alternatives provide protein from plant or vegetable sources.
- They are important for people who don't eat meat or animal products.
- There are three main protein alternatives: **soya** (e.g. textured vegetable protein (TVP) and tofu); **mycoprotein** (Quorn); and **quinoa** (a bead-shaped seed). All of these are high biological value.

Soya

- Fresh soya beans are known as edamame beans and are eaten in salads.
- Dried soya beans can be made into TVP, tofu, soya milk, tempeh and miso.

Mycoprotein

- Quorn is made from mycoprotein, a type of fungus grown under special conditions.
- It can be made into different shapes (e.g. sausages, burgers) and minced.

Quinoa

- Quinoa are tiny, bead-shaped seeds that can be used in a wide range of dishes (e.g. curries, soups, salads).
- Quinoa is an HBV protein.

Figure 1.1 Soya beans are an important high biological value protein

> **Typical mistake**
>
> Not all protein alternatives are suitable for vegetarians. Some Quorn products are made with egg white so cannot be eaten by vegans, although a Quorn vegan range has recently been launched.

Dietary reference value for protein

REVISED

- Boys aged 11–14 years require **42.1 g** of protein each day.
- Girls aged 11–14 years require **41.2 g** of protein each day.
- Men require more protein than women due to the fact they are usually bigger.
- Babies and children require a lot of protein because they are growing.
- Teenagers need more protein to support their rapid growth spurt.

Deficiency and excess of protein

REVISED

- Protein deficiency is very rare in the developed world.
- **Kwashiorkor** is a deficiency of protein and energy. Children suffering from kwashiorkor have poor growth rates and persistent infections.

> ### Now test yourself
>
> TESTED
>
> 1 List three functions of protein in the diet. [3 marks]
> 2 Explain the difference between high biological value protein and low biological value protein. [2 marks]
> 3 Name three sources of plant protein. [3 marks]
> 4 Describe two products that are made from soya beans. [4 marks]
> 5 Explain the term protein complementation. [2 marks]

Fats and oils

- All fats provide us with energy; 1 g of fat provides 9 kcals of energy.
- Fat intake should not be more than 35 per cent of total energy intake.
- Excess fat is stored as body fat.

Fats and oils have an important role in improving the flavour, texture and smell of food. They make food crispy, crumbly and moist.

Functions, sources, deficiency and excess

REVISED

Table 1.2 **Functions and sources of fats**

Functions in the body	Provide energy	
	Keep the body warm, as adipose tissue under the skin	
	Form part of every body cell	
	Protect organs (e.g. kidneys)	
	Provide the fat-soluble vitamins A, D, E and K	
	Provide the essential fatty acids	
	Make you feel full for longer because fats slow down the rate at which the stomach empties	
Sources of fat	**Animal sources**	**Vegetable sources**
	Butter, ghee	Vegetable and plant oils
	Lard, goose fat, suet, dripping	Avocados and olives
	Meat and meat products	Nuts and nut products
	Oily fish	Seeds
	Full-fat Greek yoghurt	Fat spreads
	Hard cheese	
	Cream	
	Eggs	
	Chocolate, pastries, biscuits, cakes	
What happens if we don't get enough fat?	Weight loss	
	Lack of essential fatty acids	
	Lack of vitamins A, D, E and K	
What happens if we get too much fat?	Weight gain	
	Obesity	
	Raises 'bad cholesterol' levels in the body	
	Risk of type 2 diabetes, high blood pressure and heart disease	

Types of fats and oils

REVISED

- The chemical name for a fat is a **triglyceride**.
- A triglyceride molecule is made of three fatty acid parts attached to one glycerol part.
- The fatty acids can either be:
  - ○ **saturated** (full up) with hydrogen atoms
  - ○ **unsaturated** (not full up) with hydrogen atoms.

Saturated fats

REVISED

- Saturated fats are mainly **animal foods** (e.g. red meat, butter, ghee, cream, hard cheese, eggs).
- Too much saturated fat in the diet has been linked to high blood cholesterol, which causes an increased risk of **heart disease**, **type 2 diabetes** and **obesity**.
- Only 11 per cent of our energy intake should come from saturated fat.

Unsaturated fats

REVISED

- Unsaturated fats are found in animal and plant foods (e.g. oily fish, nuts and seeds).
- Unsaturated fats are healthier than saturated fats. They may lower blood cholesterol levels and reduce the risk of heart disease.
- Monounsaturated fatty acids have **one double bond** (e.g. avocados, cashews and peanuts).
- Polyunsaturated fatty acids have **two or more double bonds** (e.g. sunflower oil).

> **Typical mistake**
>
> Only low-fat spread is low in fat. Butter, margarine and vegetable oils all contain at least 80 per cent fat.

Essential fatty acids

- Omega 3 and omega 6 are **essential fatty acids** and must be eaten in the diet as the body cannot make them. They are vital for the proper functioning of the brain and nervous system.
- **Omega 3** is found in oily fish, seeds and green leafy vegetables.
- **Omega 6** is found in vegetables, grains, seeds and chicken.

Cholesterol

Figure 1.2 Fat spread

- Cholesterol is a fatty substance that is needed by the body to make cell membranes and help with the digestion of fats.
- Eating foods that are high in saturated fat will raise cholesterol levels in the blood.
- Cholesterol is carried around the body by proteins called **lipoproteins**.
- There are two types of lipoprotein:
  - **low-density lipoprotein (LDL)**, called 'bad cholesterol'
  - **high-density lipoprotein (HDL)**, called 'good cholesterol'.
- Too much bad cholesterol and saturated fat in the body can build up in arteries and cause heart disease.
- Good cholesterol may actually help to protect against heart disease.

Now test yourself

TESTED

1 List three functions of fat in the diet. [3 marks]
2 Explain the difference between visible and invisible fat. [2 marks]
3 Name three sources of vegetable fat. [3 marks]
4 Explain the link between heart disease and fat. [4 marks]
5 Describe the role of cholesterol in the diet. [4 marks]

> **Exam tip**
>
> Accurate spelling of the technical terms cholesterol, hydrogenation, diabetes and saturated fats is important.

Carbohydrates

- Carbohydrates give the body energy.
- Carbohydrates can be divided into two groups: **sugars** and **starches**.
- Sugars are the simplest form of carbohydrates.
- There are two types of sugar: **monosaccharides** and **disaccharides**.
- Starches are more **complex carbohydrates**.
- Complex carbohydrates provide **dietary fibre**, which helps digestion.
- Sugars are absorbed quickly into the body, providing an instant burst of energy.

Functions, sources, deficiency and excess

REVISED

Table 1.3 **Functions and sources of carbohydrate**

Functions in the body	Energy for movement, growth, chemical reactions and processes	
Sources of carbohydrate	**Sugar sources**	**Starch sources**
	All types of sugar	Root vegetables (e.g. potatoes, carrots)
	Treacle and golden syrup	
	Honey, jam and marmalades	Cereals and cereal products (e.g. bread, pasta, rice, beans, breakfast cereals)
What happens if we don't get enough carbohydrate?	Lose fat and weight	
	Poor growth in children	
What happens if we get too much carbohydrate?	Increase in body fat and weight, leading to obesity	
	Too much sugar will cause tooth decay	

Types of carbohydrate

REVISED

- Carbohydrates can be divided into **simple sugars** and **complex carbohydrates**.
- The simple sugars are **monosaccharides** and **disaccharides**.
- The **complex carbohydrates** are the **polysaccharides**.

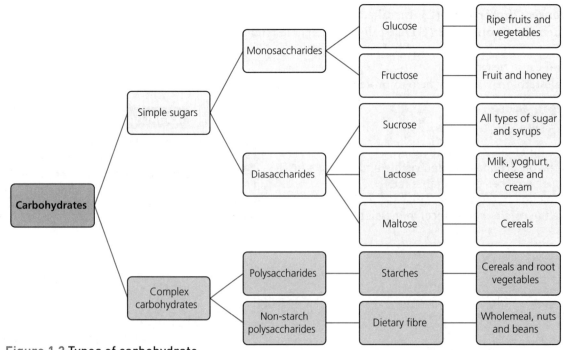

Figure 1.3 **Types of carbohydrate**

Exam practice answers and quick quizzes at **www.hoddereducation.co.uk/myrevisionnotes**

Types of sugar

REVISED

- Sugar can be described as a **free sugar** or **fruit sugar**.
- **Fruit sugars** are found naturally inside fruit and vegetable cells (e.g. sugar in fresh fruit).
- **Free sugars** are added to food or found outside the cell structure (e.g. granulated sugar, icing sugar, treacle, syrup and honey).
- A diet high in free sugars can lead to tooth decay and obesity.
- Sugar provides energy but contains no other nutrients.
- Many processed foods contain **hidden sugar**.
- Hidden sugars are found in savoury foods like salad dressings, bread, sauces and soups.
- The names of hidden sugars can be found on food labels. They are **corn sugar**, **dextrose**, **fructose** and **glucose**.

Complex carbohydrates

REVISED

- Complex carbohydrates, or **polysaccharides**, are made up of many simple sugars (glucose) joined together.
- **Starch**, **pectin** and **glycogen** are complex carbohydrates.
- **Starch** is found in cereals and root vegetables.
- **Pectin** is found naturally in some fruits and helps jams to set.
- **Glycogen** is made from glucose by humans; it is stored in the liver and muscles as an energy reserve.
- **Dietary fibre** is a complex carbohydrate.

Figure 1.4 Bran flakes are a good source of dietary fibre

Dietary reference values for sugars and starches

REVISED

- 50 per cent of total food energy should be from carbohydrates.
- 45 per cent should be from starchy carbohydrate, milk sugar and fruit sugar.
- Free sugars should be restricted to providing 5 per cent of daily energy (calorie) requirements.
- Teenagers consume 50 per cent more sugar on average than is currently recommended.

Dietary fibre

REVISED

- **Dietary fibre** is a polysaccharide found in the cell walls of vegetables, fruits, pulses and cereal grains.
- Dietary fibre cannot be broken down by the digestive system so passes through the intestine **undigested**.
- Dietary fibre helps the digestive system to work properly.
- There are two types of dietary fibre: **soluble fibre** and **insoluble fibre**.
  - **Insoluble fibre** passes through the body mostly unchanged as it is undigested.
  - **Soluble fibre** slows down the digestion and absorption of carbohydrates, so it helps to control blood sugar levels.

Exam tip

You may be asked to state the sources of dietary fibre. Always be specific and avoid generalisations (e.g. 'vegetables' is vague; instead try 'vegetables eaten with their skins').

Functions, sources, deficiency and excess

Table 1.4 Dietary fibre

Functions in the body	Allowing the digestive system to function properly	
	Helping weight control as high-fibre foods release energy slowly and leave us feeling fuller for longer	
	Preventing some bowel diseases, e.g. diverticular disease	
	Providing soluble fibre, which can help to reduce cholesterol levels	
Sources of dietary fibre	**Insoluble fibre**	**Soluble fibre**
	Wholegrain foods (e.g. wholegrain bread, breakfast cereals and pasta)	Oats
		Nuts
	Brown rice	Legumes (e.g. dried peas, beans and lentils)
	Wheat bran	
	Fruit and vegetable peel and skins	Fruits (e.g. prunes, bananas, apples, pears, plums)
	Nuts and seeds	Vegetables (e.g. potatoes, sweet potatoes, broccoli, carrots)
What happens if we don't get enough dietary fibre?	Constipation	
	Increased risk of bowel cancer	
What happens if we get too much dietary fibre?	Too much fibre can reduce the body's ability to absorb iron and calcium	

Dietary reference values for dietary fibre

- The dietary reference value (DRV) for dietary fibre is 30 g for adults.
- Children should eat less because of their small body size.
- Very young children should avoid too many fibre-rich foods as being full up with fibre can make it difficult for them to meet their other nutritional needs.

Typical mistake

Some students assume that everyone should eat more dietary fibre. Remember that very young children should avoid too many fibre-rich foods because they can slow down the absorption of some nutrients.

Now test yourself

1 Name three processes that require energy in the body. [3 marks]
2 Explain the difference between a free sugar and a fruit sugar. [2 marks]
3 Explain the difference between soluble and insoluble dietary fibre. [2 marks]
4 Name two complex carbohydrates. [2 marks]

2 Micronutrients

Micronutrients are needed by the body in small amounts. Vitamins and minerals are micronutrients.

Vitamins

There are two types of vitamins: fat-soluble vitamins and water-soluble vitamins.

Fat-soluble vitamins

REVISED

- Vitamins A, D, E and K are **fat-soluble vitamins**.
- All fat-soluble vitamins are stored in the **liver**.

> **Exam tip**
>
> The fat-soluble vitamins are **KADE**:
>   vitamin **K**
>   vitamin **A**
>   vitamin **D**
>   vitamin **E**

Table 2.1 Functions and sources of fat-soluble vitamins

Vitamin	Functions in the body	Sources	
A	Growth Helps vision in dim light Keeps the skin healthy Protects the body; it is an **antioxidant**	**Retinol** (animal sources): eggs, oily fish, liver, full-fat milk, butter and cheese, fortified margarines and fat spreads	**Beta-carotene** (vegetable sources): spinach, carrots, sweet potatoes, red peppers, mangoes, apricots
D	Prevents bone diseases Helps the body to absorb calcium Develops and maintains bones and teeth Heals broken bones	sunshine, milk, butter, liver, oily fish and eggs, fortified breakfast cereals, fat spreads	
E	Protects the body; it is an antioxidant Forms red blood cells	green peas, green beans, broccoli, spinach, vegetable oils, cereals	
K	Helps blood to clot Maintains bone health	green leafy vegetables (e.g. broccoli, spinach), vegetable oils, cereal grains	

Table 2.2 Dietary reference values of the fat-soluble vitamins for 11–14 year olds

Fat-soluble vitamins	Male	Female
A	600 mcg	600 mcg
D	10 mcg	10 mcg
E	4 mg	3 mg
K	0.045 mg	0.045 mg

Table 2.3 Deficiency and excess of fat-soluble vitamins

Vitamin	What happens if we don't get enough?	What happens if we get too much?
A	Night blindness	Poisonous if eaten in large amounts Pregnant women should avoid foods high in vitamin A
D	**Rickets** in babies and toddlers **Osteoporosis** in adults People who are not exposed to much sun and people who are of African, African-Caribbean and South Asian origin can suffer from a lack of vitamin D	Rare
E	Very rare	Loss of appetite
K	Blood may take longer to clot A very small number of babies suffer bleeding due to a lack of vitamin K	Rare

Water-soluble vitamins

REVISED

- The B group of vitamins and vitamin C are water soluble.
- **Water-soluble vitamins** cannot be stored in the body and need to be eaten regularly.

Exam tip

'Blue sea' (BC) are the water-soluble vitamins B and C.

Table 2.4 **Functions and sources** of water-soluble vitamins

Vitamin	Functions in the body	Sources
B1 (thiamin)	Releases energy from food Helps the nervous system	Liver, milk and cheese, bread, fortified breakfast cereals, dried fruit, eggs, potatoes, nuts, peas
B2 (riboflavin)	Helps the body to release energy Keeps skin, eyes, nerves and body tissue healthy	Chicken, eggs, milk, fish, yoghurt, leafy vegetables, rice, bread, breakfast cereals, soya beans
B3 (niacin)	Helps to release energy from food Keeps skin and nerves healthy	Meat, fish, flour, eggs, milk
Folic acid	Reduces the risk of nervous system faults in unborn babies Works with vitamin B12 to make blood	Fortified breakfast cereals, broccoli, brussels sprouts, liver, chickpeas, spinach, asparagus, peas
B12	Maintains nerves Makes blood Releases energy	Meat, eggs, milk and cheese, salmon and cod, fortified breakfast cereals

Vitamin	Functions in the body	Sources
C	Makes and maintains healthy connective tissue Helps wounds to heal Helps the absorption of iron Protects the body; it is an antioxidant	Oranges and orange juice, blackcurrants, broccoli, potatoes, red and green peppers, strawberries, brussels sprouts, tomatoes

Table 2.5 Deficiency and excess of water-soluble vitamins

Vitamin	What happens if we don't get enough?	What happens if we get too much?
B1 (thiamin)	Beriberi	No side effects
B2 (riboflavin)	Rare	No side effects
B3 (niacin)	Pellagra	Liver damage
Folic acid	**Spina bifida** Before pregnancy a woman should take 300 mcg a day	No side effects
B12	**Pernicious anaemia** Vegans do not eat any animal products so are at risk	No side effects
C	**Scurvy**	Stomach pain Diarrhoea

Table 2.6 Dietary reference values of the water-soluble vitamins for 11–14 year olds

Fat-soluble vitamins	Male	Female
B1	0.9 mg	0.7 mg
B2	15 mg	12 mg
B3	15 mg	12 mg
Folic acid	200 mcg	200 mcg
B12	1.2 mcg	1.2 mcg
C	35 mg	35 mg

Revision activity

Draw a web diagram on a large piece of paper, using coloured pens, to show:
- functions in the body of the vitamins
- sources
- what happens if we don't get enough
- what happens if we get too much
- DRVs of the vitamins for teenagers.

Exam tip

Be specific about the functions of each vitamin in the exam. Avoid general statements (e.g. 'protect the body' or 'keep you healthy')

Figure 2.1 **Sources of water-soluble vitamins**

Buying, storing, preparing and cooking fruit and vegetables

Water-soluble vitamins (B group and vitamin C) dissolve in water, can be destroyed by contact with sunlight, air and heat, and are affected by **enzymes**.

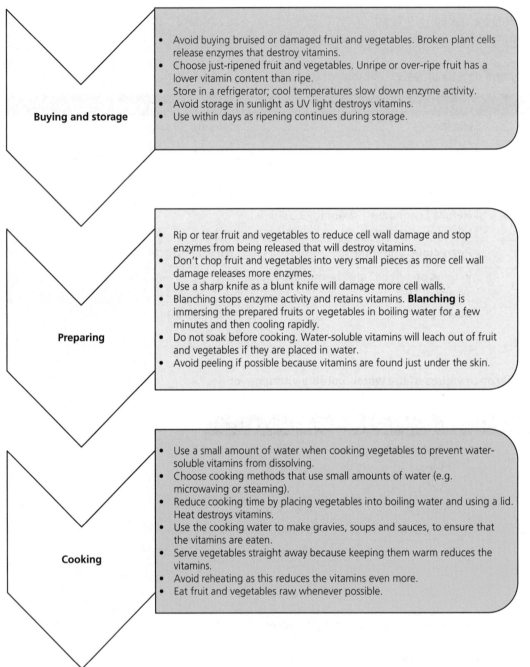

Buying and storage
- Avoid buying bruised or damaged fruit and vegetables. Broken plant cells release enzymes that destroy vitamins.
- Choose just-ripened fruit and vegetables. Unripe or over-ripe fruit has a lower vitamin content than ripe.
- Store in a refrigerator; cool temperatures slow down enzyme activity.
- Avoid storage in sunlight as UV light destroys vitamins.
- Use within days as ripening continues during storage.

Preparing
- Rip or tear fruit and vegetables to reduce cell wall damage and stop enzymes from being released that will destroy vitamins.
- Don't chop fruit and vegetables into very small pieces as more cell wall damage releases more enzymes.
- Use a sharp knife as a blunt knife will damage more cell walls.
- Blanching stops enzyme activity and retains vitamins. **Blanching** is immersing the prepared fruits or vegetables in boiling water for a few minutes and then cooling rapidly.
- Do not soak before cooking. Water-soluble vitamins will leach out of fruit and vegetables if they are placed in water.
- Avoid peeling if possible because vitamins are found just under the skin.

Cooking
- Use a small amount of water when cooking vegetables to prevent water-soluble vitamins from dissolving.
- Choose cooking methods that use small amounts of water (e.g. microwaving or steaming).
- Reduce cooking time by placing vegetables into boiling water and using a lid. Heat destroys vitamins.
- Use the cooking water to make gravies, soups and sauces, to ensure that the vitamins are eaten.
- Serve vegetables straight away because keeping them warm reduces the vitamins.
- Avoid reheating as this reduces the vitamins even more.
- Eat fruit and vegetables raw whenever possible.

Figure 2.2 Preventing vitamin loss in fruit and vegetables

Typical mistake

There is no loss of fat-soluble vitamins (A, D, E and K) in water. Only B vitamins and vitamin C are water soluble.

Antioxidants

- All bodily functions and lifestyle habits produce substances called **free radicals**.
- Free radicals attack healthy cells, and increase the risk of heart disease and certain types of cancer.
- **Antioxidants** protect healthy cells from the damage caused by free radicals.
- **Vitamin A**, **vitamin C** and **vitamin E** are all antioxidants.
- There are large amounts of antioxidants in fruits, vegetables, nuts and whole grains, and smaller amounts of antioxidants in meat, chicken and fish.

Now test yourself

1 Name two sources of retinol. [2 marks]
2 State what happens if you don't get enough vitamin C. [1 mark]
3 List three functions of vitamin D in the diet. [3 marks]
4 Explain why vegans may have a problem getting enough vitamin B12. [2 marks]
5 Describe four ways you can make sure water-soluble vitamins are retained during food preparation. [4 marks]
6 Name three vitamins that are antioxidants. [3 marks]

Minerals

Sources and functions of minerals

Table 2.7 Sources and functions of minerals

Mineral	Functions in the body	Sources	
Calcium	Builds strong bones and teeth Controls muscle function Controls heartbeat Helps blood clotting	Nuts, bread and fortified cereals, cheese, milk, green leafy vegetables, oily fish, soya and tofu	
Iron	Makes haemoglobin in red blood cells Carries oxygen around the body	**Animal sources** (haem iron): liver, meat, eggs	**Plant sources** (non-haem iron): fortified cereals and bread, green leafy vegetables, nuts, dried fruit
Sodium (salt)	Maintains water balance in the body	Cheese, salted nuts, smoked fish, bacon, bread, crisps, ready meals, tinned foods	
Fluoride	Prevents tooth decay Supports bone health	Drinking water (fluoride is added to the water supply in some areas through fluoridation), sardines, seafood, tea	
Iodine	Makes the hormone thyroxine Maintains a healthy metabolic rate	Red meat, sea fish and shellfish, cereals and grains	
Phosphorus	Maintains bones and teeth with calcium Releases energy from food	Red meat, dairy foods, fish, poultry, bread, brown rice, oats	

Table 2.8 Deficiency and excess of minerals

Mineral	What happens if we don't get enough?	What happens if we get too much?
Calcium	Rickets in children Osteoporosis in adults	Stomach pain Diarrhoea
Iron	Iron deficiency anaemia (more common in girls than boys) Shortness of breath A pale appearance Brittle nails Cracked lips	Constipation Feeling sick Stomach pain
Sodium (salt)	(Iron works with vitamin C to prevent iron deficiency anaemia) Muscle cramps	Increase in blood pressure Stroke Heart attack
Fluoride	Tooth decay	Staining and pits develop on the teeth
Iodine	Goitre	Rare
Phosphorus	Increased risk of bone fractures	Rare

Table 2.9 Dietary reference values of the key minerals for 11–14 year olds

Minerals	Male	Female
Calcium	1,000 mg	800 mg
Iron	11.3 mg	14.8 mg
Salt	6 g	6 g
Fluoride	2 mg	2 mg
Iodine	130 mcg	130 mcg
Phosphorus	775 mg	625 mg

Typical mistake

Minerals do not provide energy.

Exam tip

Be specific in your answers. Calcium is required for strong bones and strong teeth. Do not use the term 'healthy' bones.

Revision activity

Draw a web diagram on a large piece of paper, using coloured pens to show:
● functions in the body of minerals
● sources
● what happens if we don't get enough
● what happens if we get too much
● DRVs of the key minerals for teenagers.

Exam tip

Try to remember to use the correct unit when mentioning the DRVs.

The amount of calcium, iron, sodium, fluoride and phosphorus you need each day is measured in **milligrams (mg)**.

The amount of iodine you need is measured in **micrograms (mcg)**.

Figure 2.3 Osteoporosis is linked to a low intake of calcium during childhood

Now test yourself

TESTED ☐

1 Name four sources of calcium. [4 marks]
2 Describe one function of iron in the diet. [1 mark]
3 Name two foods that provide high amounts of salt. [2 marks]
4 Describe the effects of too much fluoride in the diet. [2 marks]
5 Explain two reasons why phosphorous is important in the diet. [4 marks]

Water

- The human body can only survive a few days without water.
- **Hydration** is the supply of water required to maintain the correct amount of fluid in your body.
- **Dehydration** occurs when your body loses more water than you take in.

Functions of water

REVISED

- Cools the body by sweating, to prevent cell damage and overheating
- Removes waste from the body
- Transports waste products from the body

Figure 2.5 Foods that contain water

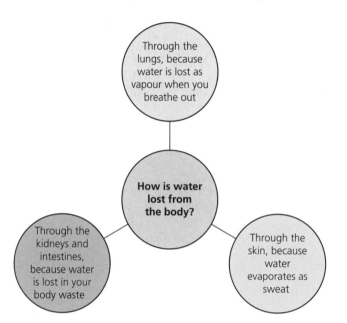

Figure 2.4 How is water lost from the body?

- Water is an important part of **saliva**, which you need in order to be able to swallow food.
- Water is an important party of **digestion** because, in the stomach, food mixes with acid and water.
- Water is an important way of moving food around the body because the bloodstream is about 90 per cent water.
- Most people need about **2 litres of water a day**, which is about eight average-size glasses.
- Water can also be found in low-sugar soft drinks, milk, fruit juices, drinks made with water (for example, tea, coffee and fruits).
- If we do not get enough water we feel thirsty and become dehydrated. You may have dark urine, less urine, experience a headache, lack of energy or light-headedness.

> **Exam tip**
>
> Learn a key fact: the amount of water required each day.

Table 2.10 Factors that affect the amount of water you need

Health	Fever can increase water loss
	Vomiting and diarrhoea can cause high losses of fluids
Age	Young children need lots of water relative to their size because they are very active and have a higher surface area of skin compared to adults
	Older people may have a weaker sense of thirst and, if necessary, should be reminded to drink regularly
Gender	Generally men are slightly bigger than women so require more water each day
Physical activity	During exercise the body sweats to cool down
	The longer the exercise, the greater the demand to replace lost fluid
Environment	Hot or humid weather increases sweat losses
	Heated indoor air can increase sweat and skin losses during the winter
Breastfeeding	Women who are breastfeeding require extra water to produce milk
Eating salty foods	Salt makes body fluids more concentrated
	This makes us thirsty and we need more water until the excess salt has been removed by the kidneys

Typical mistake

Don't assume that all drinks are good for you.

It is important to be aware that some drinks are acidic (e.g. fruit juice and fizzy drinks) and full of sugar. This may cause tooth decay if they are consumed frequently.

Now test yourself

TESTED

1 Describe two functions of water in the diet. [4 marks]
2 How much water is needed each day? [1 mark]
3 Describe two effects of a lack of water on the body. [4 marks]
4 Explain four factors that can affect the amount of water required. [8 marks]

Revision activity

Draw a web diagram showing all the different factors that affect our water requirements.

3 Nutritional needs and health

Making informed choices about a varied and balanced diet

- A healthy diet is low in fat, salt and sugar, and high in fibre.
- A **balanced diet** contains all the required nutrients in the correct amounts to meet individual needs.
- The **Eatwell Guide** shows the proportions in which different groups of foods can contribute to a healthy balanced diet.

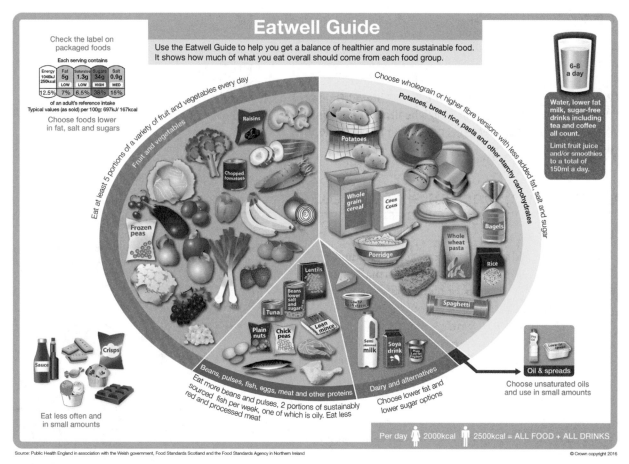

Figure 3.1 **The Eatwell Guide**

What are the current guidelines for a healthy diet?

REVISED

To help you choose a healthy diet, Public Health England has produced the following eight tips for eating well.

1 Base your meals on starchy foods.
2 Eat lots of fruit and veg.
3 Eat more fish – including a portion of oily fish each week.
4 Cut down on saturated fat and sugar.
5 Eat less salt – no more than 6 g a day.

6 Get active and be a healthy weight.

7 Don't get thirsty.

8 Don't skip breakfast.

Remember that fruit juice and/or smoothies should be limited to no more than 150 ml per day in total.

Typical mistake

You may be asked to describe a healthy diet. A common exam mistake is to make vague comments such as 'choosing healthy foods' or 'having a balanced diet'. You will gain more marks if you write about:

● choosing wholegrain cereals (e.g. wholemeal bread and brown rice) as these are higher in fibre
● following the proportions of the Eatwell Guide when planning meals and diets
● including fish in your diet at least twice a week.

Exam tip

You will need to learn the eight tips for eating well and be able to give examples for each one.

Now test yourself

TESTED

1 Give a concise definition of:
  (a) a healthy diet   [2 marks]
  (b) a balanced diet.   [2 marks]
2 Which food group is the largest on the Eatwell Guide?   [1 mark]
3 Why is taking exercise and staying a healthy weight important?   [2 marks]
4 Why are wholegrain cereals recommended for a healthy diet?   [2 marks]

Meal planning: portion size and cost

REVISED

Portion size

● Portion size should be adjusted for each individual. Younger children and the elderly will normally need smaller portions than adults or teenagers.
● Portion sizes that are too large encourage overeating, which can lead to weight gain.
● In recent years portion sizes have become too large.
● One way to control portion sizes is to use smaller plates.

Cost of food

● Food prices have increased in recent years and become a larger part of the family budget.
● Students living away from home and the elderly are more likely to be on limited incomes.
● Ways to save money on food:
  ○ plan meals in advance to help prevent food waste
  ○ price comparison websites allow comparisons of food prices to be made
  ○ write a shopping list and stick to it
  ○ look out for special offers such as buy one get one free (BOGOF)
  ○ supermarket value lines can save you money
  ○ don't shop when you are hungry as you are more likely to impulse-buy

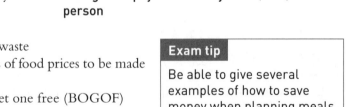

Figure 3.2 Adjust portion sizes depending on the age and physical activity level (PAL) of each person

Exam tip

Be able to give several examples of how to save money when planning meals and shopping for food.

- shop towards the end of the day when price reductions are made on items close to their date mark
- food shopping online can be easier for families with children, to avoid 'pester power' and impulse buying.

Now test yourself

TESTED

1 Name two groups of people who would need smaller meal portion sizes.   [2 marks]
2 Give two reasons why large portion sizes can be a problem.   [2 marks]
3 Give three ways a family with children can reduce its food bills.   [3 marks]

Typical mistake

Many students just mention 'buy one get one free' when asked about saving money when shopping. This offer may not be suitable for students living away from home or the elderly as it may lead to food waste.

How people's nutritional needs change, and planning for different life stages

REVISED

Young children (aged 1–4 years)

- Young children are growing rapidly and are usually very active.
- Their stomachs are small, so they cannot eat large meals
- They need regular, smaller meals as well as snacks and drinks throughout the day to provide all their nutritional needs and enough energy.
- The Eatwell Guide doesn't apply to children under 2 years, but their diets should gradually be adapted between the ages of 2 and 5 years.
- Young children benefit from having whole milk in their diet as it provides more energy as well as more vitamin D.
- Vitamin D allows the proper absorption of calcium for strong bones and teeth.

Figure 3.3 Whole milk supplies vitamin D, which is needed to absorb calcium properly

School children (aged 5–12 years)

- School children are growing fast and most are active.
- Some school children are not as active as they once were.
- They should have a varied and balanced diet following the Eatwell Guide.
- The consumption of fatty and sugary foods has increased. Over a fifth of 4–5 year olds and a third of 10–11 year olds were overweight in 2014/15.
- It is better for school children to have school meals rather than a packed lunch as these provide a well-balanced meal.

- The School Food Plan sets out standards for all food served in most state schools.
- Water is the best drink for school children as is doesn't contain free sugars.
- Snacks such as fresh fruit and starchy carbohydrate (e.g. bread, plain crackers) are best.

Teenagers

- Teenagers should follow the guidelines of the Eatwell Guide.
- Teenagers have rapid growth spurts. Over a third of teenagers are overweight, so an increase in kilocalories (food energy) may not be needed for these growth spurts.
- Boys need more protein than girls as they develop more muscle tissue and are usually a bigger build. They also need more calcium than girls for their higher bone mass.
- To be able to absorb calcium properly, both boys and girls need enough vitamin D to form the skeleton.
- The skeleton is not fully formed until the late twenties, so the teenage years are important to enable peak bone mass to be reached.
- Girls need more iron than boys due to menstruation.
- Breakfast is an important meal and aids concentration.
- Regular meals should be eaten to control blood sugar levels.
- The diet should be as natural as possible, avoiding processed foods.
- Wholegrain cereals are best as these increase the fibre in the diet.
- Avoid sugary fizzy drinks as these contain very high sugar levels and few other nutrients.

Figure 3.4 Many fizzy drinks are high in sugar

Adults

- Adults need a well-balanced diet following the Eatwell Guide.
- The amount of energy (kilocalories) consumed should depend on the physical activity level to achieve a healthy body weight.
- Avoid excess sugar to reduce the risk of dental decay and to reduce energy intake.
- A poor diet may lead to conditions such as coronary heart disease (CHD), some cancers and obesity. More than half of adults in the UK are overweight or obese.
- Calcium and vitamin D are important to ensure bones stay strong. Iron and vitamin C are important to keep the red blood cells healthy. Women of child-bearing age require more iron than men.

The elderly (over 65)

- Elderly people still need to follow the Eatwell Guide.
- A well-balanced diet containing all of the nutrients in the correct proportions is important.
- Less energy (fewer kilocalories) is needed as the bodies of the elderly tend to slow down and become less efficient than they were.
- Sugary and fatty foods should be avoided as these are energy dense, providing excessive energy.
- Elderly people may develop weak bones. This is known as **osteoporosis** and can be a problem in old age, which is why both calcium and vitamin D are important.
- Exercise also improves bone strength as it improves the take-up of calcium by the bones.
- Dietary fibre and sufficient liquids are important to ensure the proper working of the digestive system to prevent constipation, diverticular disease and cancer of the bowel.
- Iron deficiency anaemia may be prevented by consuming enough iron and vitamin C. Vitamin C helps the absorption of iron.
- High blood pressure can be reduced by lowering the amount of salt in the diet. Elderly people should restrict their consumption of ready meals, and add less or no salt to meals.
- Vitamin B12 is not as easily absorbed as the body ages and some people may need injections to boost their levels of this vitamin.

Figure 3.5 A healthy diet and regular exercise slow down the effects of ageing

> **Exam tip**
>
> The Eatwell Guide applies to all target groups except children under 2 years. Make sure you learn the names of each segment of the Eatwell Guide and some examples of foods from each.

> **Typical mistake**
>
> Students can write too generally about the different target groups. Each group has its own specific needs. For example, it is wrong to generalise and say that teenage boys have higher nutritional needs than teenage girls, as girls need more iron than boys due to menstruation.

Now test yourself

TESTED

1 By what age should a child's diet be adjusted to follow the Eatwell Guide? [1 mark]
2 Which two nutrients are needed in the correct amounts to enable 'peak bone mass' to be reached? [2 marks]
3 What proportion of adults were overweight or obese in the UK in 2014/15? [1 mark]
4 Why are fibre and sufficient liquids so important for the elderly? [3 marks]

Planning balanced meals for different dietary groups

REVISED

Vegetarians

People may become vegetarian for different reasons:
- they do not like the idea of eating a dead animal, fish or bird
- they think it is cruel to kill animals for food
- religious reasons
- health reasons
- economic reasons (wasteful to raise animals when the same land space could grow many more crops).

There are three main types of vegetarians.

Lacto vegetarians

Lacto vegetarians:
- **will** eat dairy products
- **will not** eat meat, poultry and fish
- **will not** eat products made from animals, such as lard and gelatine
- **will not** eat eggs.

Lacto-ovo vegetarians

Lacto-ovo vegetarians:
- **will** eat eggs
- **will** eat dairy products
- **will not** eat products made from animals, such as lard and gelatine
- **will not** eat meat, poultry and fish.

Nutritional deficiencies are unlikely when following a lacto or lacto-ovo vegetarian diet but iron could possibly become deficient, as the iron from non-meat foods (non-haem iron) is more difficult for the body to absorb.

Vegans

- A vegan is sometimes called a 'strict vegetarian'.
- A vegan diet contains no animal foods. All foods eaten by a vegan are plant based.
- No food that involves the use of animals in its production is included in the diet. This means no dairy foods or honey are included in a vegan diet.

Some possible deficiencies arising from a vegan diet are as follows.
- Nutritional deficiencies are more likely when following a vegan diet, unless it is carefully planned.
- The following nutrients could become deficient in a vegan diet: protein, calcium, iron, vitamin A and vitamin B12.
- Vitamin B12 is not found in plant foods, so foods fortified with this vitamin should be consumed (e.g. soya milk, almond milk and some breakfast cereals).

> **Exam tip**
>
> Have an understanding of the different types of vegetarians and what they may include in their diets. Be able to plan a balanced meal for a vegetarian or vegan.

Figure 3.6 Foods fortified with vitamin B12 and suitable for vegans

Exam practice answers and quick quizzes at **www.hoddereducation.co.uk/myrevisionnotes**

Coeliac disease

- **Coeliac disease** is a sensitivity to gluten. Gluten is a protein found in wheat, rye, barley and sometimes oats.
- People with coeliac disease react to gluten when eaten and their body attacks its own healthy tissue.
- This can damage the lining of the intestine and stops nutrients from being absorbed.
- Food labelling laws mean that gluten must be listed on the label if present.
- To ensure coeliacs have a balanced diet, they should obtain starchy carbohydrates from gluten-free cereals such as rice and gluten-free oats.
- Supermarkets have extended their range of gluten-free products and often have a separate aisle for them.

Lactose intolerance

- **Lactose intolerance** means the body cannot digest lactose, which is the sugar in milk.
- People with lactose intolerance don't produce enough of an enzyme called lactase, which breaks down lactose (a disaccharide) into the monosaccharides: glucose and galactose.
- This condition may be temporary or permanent.
- To ensure a balanced diet, milk and dairy substitute products are available such as lactose-free milk as well as alterative milks such as soya and almond milks.

High-fibre diet

There are two different types of fibre: **soluble fibre** and **insoluble fibre**.

- ○ **Soluble fibre** can be digested. It is found in cereals such as oats, barley and rye, fruits such as apples and bananas, and most root vegetables such as carrots and potatoes. Soluble fibre helps to prevent constipation and can also help to reduce cholesterol.
- ○ **Insoluble fibre** cannot be digested. It helps waste food to pass out of the digestive system more easily and also helps to prevent constipation. Insoluble fibre is found in wholemeal bread, wholegrain cereals, nuts and seeds.
- ○ A benefit of a high-fibre diet is that it helps to fill you up so you are less likely to overeat and become overweight.
- ○ Increasing the amount of wholegrain cereals and fruits and vegetables in the diet is likely to improve the balance of the diet as these are the largest segments of the Eatwell Guide.

Figure 3.7 Bread products containing gluten should be avoided by people with coeliac disease

> **Typical mistake**
>
> Students can sometimes focus too much on what the different dietary groups cannot eat, rather than what they can eat. Coeliacs cannot eat wheat flour, but there are other flour blends that are successful in baking containing, for example, rice flour and potato flour.

> **Revision activity**
>
> Copy out and complete the table below using the information above.
>
> | Type of fibre | Food sources | Benefits to health |
> | --- | --- | --- |
> | Soluble fibre | | |
> | Insoluble fibre | | |

Now test yourself

TESTED ☐

1 Which two groups of vegetarians do not eat eggs? [2 marks]
2 Name two foods a coeliac should avoid. [2 marks]
3 Give three reasons why a high-fibre diet is beneficial to health. [3 marks]

Energy needs

Your body needs energy for every function and movement it performs, for example:

- breathing
- the function of internal organs and for digesting food
- activities such as walking, running, cycling and even sitting down.

How is energy measured?

You get energy from the food you eat. This is measured in kilocalories (kcal) or kilojoules (kj).

- Kilocalories is the unit used most often.
- A kilocalorie is the amount of heat energy needed to raise the temperature of 1g of pure water by 1°C.

Sources of energy

Table 3.1 **Sources of energy**

1 g of each nutrient	Energy value in kilocalories (kcal)
Protein	4.0
Fat	9.0
Carbohydrate	3.75

- Some foods may be described as **energy dense**. This means that, for a given weight, they contain a lot of kilocalories. Energy-dense foods are usually high in fat and sugar.
- The opposite of energy-dense food is low-calorie or low-energy food. Low-energy foods are usually high in water.

Revision activity

Complete the table below.

Low-energy foods	1
	2
	3
Energy-dense foods	1
	2
	3

- Starchy carbohydrates, such as bread, rice, pasta, potatoes, chapattis and couscous, should be the main source of your body's energy.
- Fat is the most concentrated source of energy. It is found in foods such as cooking oils, cream and oily fish.
- Protein is found in foods such as meat, fish, eggs, cheese, pulses and nuts.
- Your body uses the nutrients carbohydrate and fat first for energy, but if there are not enough of these nutrients in your diet it can use protein. This is why protein is sometimes called a secondary source of energy.
- Your body breaks down starch, fat and protein into glucose before they can be used as a source of energy.

Which factors affect energy needs?

- **Age:** energy needs increase as children grow older; this peaks in the teenage years and then reduces into adulthood and old age.
- **Activity:** your energy needs will change depending on the activities you do.
- **Health:** if you are ill and cannot move around as much as usual, your energy needs will decrease. Sometimes illnesses can increase your need for energy (e.g. if your body needs to repair itself).
- **Gender:** males normally need more energy than females. This is because males are usually bigger and have a higher proportion of muscle tissue than females.

What is basal metabolic rate?

- **Basal metabolic rate (BMR)** is how much energy (kilocalories) you need to stay alive for 24 hours when warm and resting.
- Your BMR will depend on your age, size, gender and activity levels.
- Your BMR is worked out from the amount of oxygen your body needs to be able to function.

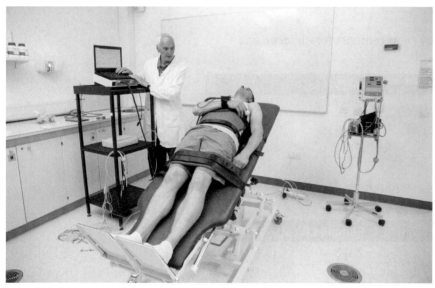

Figure 3.8 **BMR is measured when the body is warm and resting**

What is physical activity level?

- **Physical activity level (PAL)** shows your daily activity level as a number.
- If you are not very active (sedentary) you will have a lower PAL than someone more active.
- Most people have a PAL between 1.4 (a very inactive person) to 2.4 (a very active person).
- The PAL and the BMR can be used to work out how much food energy you need to consume in order to maintain your lifestyle.

$$\text{Physical activity level} = \frac{\text{Total energy expenditure (over 24 hours)}}{\text{Basal metabolic rate (over 24 hours)}}$$

Recommended energy sources from nutrients in the UK

Table 3.2 Recommended energy sources from nutrients in the UK

Protein	15% of total food energy from proteins	
Fat	No more than 35% of total food energy from fats	No more than 11% from saturated fats
Carbohydrate	50% of total food energy from carbohydrates	45% from fibre-rich starchy carbohydrate and lactose (the sugar in milk) when naturally present in milk and milk products, and sugars contained within the cellular structure of fruits and vegetables
		No more than 5% from free sugars

- Young children and teenagers need more energy in relation to their size when compared to adults and the elderly.
- After the age of about 18, energy needs begin to fall as growth has usually stopped by then.
- For the elderly, energy needs are even lower due to a reduction in their BMR and PAL.

Table 3.3 Estimated average energy requirements for children and teenagers

Age	Boys (kcal)	Girls (kcal)
4	1,386	1,291
10	2,032	1,936
18	3,155	2,462

Table 3.4 Estimated average energy requirements for adults and the elderly

Age	Males (kcal)	Females (kcal)
25–34	2,749	2,175
45–54	2,581	2,103
65–74	2,342	1,912

Exam tip

Remember: the factors that affect energy needs will affect food choices and meal planning. You should be able to describe why people's energy requirements differ.

Typical mistake

When answering questions on energy in the diet, many students wrongly state that carbohydrates provide the most energy per gram. In fact, fats provide more than double the amount of energy compared to carbohydrates per gram.

Now test yourself

TESTED

1 State three reasons your body needs energy. [3 marks]
2 Which type of carbohydrate should be the body's main source of energy? [1 mark]
3 Describe three factors that affect energy needs. [3 marks]
4 Explain what PAL means and how it is calculated. [4 marks]

How to carry out a nutritional analysis

- A nutritional analysis lets you find out the nutrients in a recipe, meal or diet.
- You can carry out nutritional analysis by using a book or website, or specially designed software.
- Nutritional analysis can help you decide if the dish, meal or diet is suitable for your target group.
- Some software will compare the nutrients with an individual's dietary reference values (DRVs). This allows you to see if your target group is having too much or too little of a particular nutrient.

Using food tables

REVISED

- Food tables can be used to find out the nutritional value of foods, a recipe or a meal.
- Foods are usually listed alphabetically and show the amount of nutrient per 100 g. This method can be very time consuming, although using spreadsheets can help to speed up this process.

Table 3.5 Nutritional analysis of a recipe for blueberry pancakes

Ingredient	Amount of iron in 100 g of ingredient	Amount of ingredient in recipe	Amount of iron in ingredient used
Plain flour	1.94 mg	100 g	1.94 × 1 = 1.94 mg
Eggs	1.72 mg	100 g	1.72 × 1 = 1.72 mg
Semi-skimmed milk	0.02 mg	300 g	0.02 × 3 = 0.06 mg
Sunflower oil	0.1 mg	20 g	0.1 × 0.2 = 0.02 mg
Blueberries	0.55 mg	150 g	0.55 × 1.5 = 0.83 mg
Total iron in recipe (serves 4)			**4.57 mg**
Total iron in one serving			**1.14 mg**

Nutritional analysis using computer software

REVISED

- Using computer software to analyse the nutrients in a recipe saves a lot of time.
- You enter the names and weights of foods used and it will produce a chart or table to show the nutrients provided.
- Some programs may compare the nutrients in a recipe with the amount an individual should be consuming (their DRVs).

Exam tip

A nutritional analysis lets you see which nutrients are in a dish/meal or diet. You will be able to gain more marks by discussing or analysing how the analysis meets the needs of a target group (e.g. a dish high in iron would be suitable for someone with iron deficiency anaemia).

Planning and modifying recipes

REVISED

- You may need to adapt a recipe to improve its nutritional value and to help meet the healthy eating guidelines.
- To make recipes healthier, you will need to make them lower in sugar, fat and salt, and higher in fibre.

Planning and modifying meals

REVISED

- Meals should be planned following the food groups and proportions of the Eatwell Guide.
- Even though individuals have different nutritional needs, varying portion sizes will allow for differences in these needs.
- Swapping some ingredients for others can make meals more suitable for target groups.

Figure 3.9 Lasagne and salad: this meal would be more suitable for an adult

Figure 3.10 Lasagne with potato wedges and salad: this meal would be more suitable for an active teenager

Planning and modifying diets

REVISED

- Your diet is made up of all the meals, snacks and drinks you consume over a longer period of time (e.g. a week or a month).
- Your diet will determine your body weight, growth and health.
- If an individual has a health problem such as obesity, their diet will need to be modified to reduce energy (kilocalories).
- If tooth decay is present, the diet will need to be modified to reduce free sugars.
- If coeliac disease is the problem, a gluten-free diet will be necessary.

Now test yourself

TESTED

1 Why is it important to eat a wide range of foods? [1 mark]
2 Which are the two main methods of carrying out nutritional analysis? [2 marks]
3 Why is nutritional analysis useful? [1 mark]
4 Why would it be useful to compare the nutrients provided in a recipe to the DRVs of a target group? [2 marks]

Diet, nutrition and health

- The food you eat affects how healthy you are.
- You should follow the principles of the Eatwell Guide and should not overeat or undereat so you can achieve a healthy balanced diet.
- **Over-nutrition** means eating too much food or too much of a nutrient.
- **Under-nutrition** means eating too little food to meet dietary needs.
- In the UK, over-nutrition is more of a problem than under-nutrition.

Obesity

REVISED

- Obesity means being very overweight.
- About one in four UK adults and one in five UK children are obese.
- School children are not as active as they once were, due to the increased use of technology; they are playing outside less and sitting down more.
- Your body mass index (BMI) can be used to see if your weight falls into the healthy range.
- You need to know your height and weight to calculate your BMI:

$$\text{Body mass index} = \frac{\text{Weight in kilograms}}{\text{Height (in metres squared)}}$$

The formula above is the calculation to work out your BMI.

- Another measure of obesity is your waist measurement. Adult men have a higher health risk if their waist is above 94 cm and women above 80 cm.

Health problems linked to obesity are:
- type 2 diabetes
- coronary heart disease
- stroke
- breast and bowel cancer
- arthritis
- depression.

Exercise is important in treating or preventing obesity – children should exercise for at least an hour a day, adults about four hours a week.

Losing weight gradually is better than fast weight loss as you are more likely to keep the weight off. Weight will be lost when the amount of kilocalories consumed is less than those expended in day-to-day activities.

Cardiovascular disease

REVISED

- Cardiovascular disease covers a group of diseases including coronary heart disease and stroke.
- If blood flow is reduced or stopped by a blood clot or narrowing of the blood vessels, damage may be caused to the body.
- If this happens in the heart; it can cause a heart attack; if this happens in the brain, the person will have a stroke.

Coronary heart disease

- Coronary heart disease occurs when blood vessels to the heart become blocked with fatty deposits.
- This can cause **angina** if the blood flow is restricted, or a **heart attack** if the blood supply is cut off completely. It is the main cause of death in the UK.

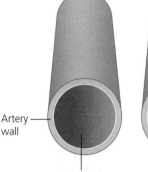

Artery wall

Blood within the artery

Fatty deposits building up

Fatty deposits develop, restricting the blood flow through the artery

Figure 3.11 How fatty deposits build up in the blood vessels

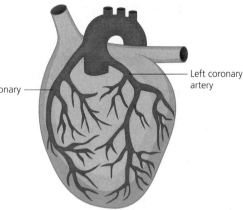

Right coronary artery

Left coronary artery

Figure 3.12 If the coronary arteries become blocked with fatty substances, it can cause a heart attack

Stroke

- A stroke occurs when the blood supply to the brain is cut off.
- A stroke may lead to physical disability, brain injury or death.
- It is the third largest cause of death in the UK after heart disease and cancer.

Reducing the risk of cardiovascular disease

- Follow the Eatwell Guide and do not overeat.
- People should not drink too much alcohol – no more than 14 units a week for men and women spread over three or more days.
- Children should learn good eating habits and take regular exercise, ideally learning from their parents.
- No more than 35 per cent of the total energy in your diet should come from fat.
- Reduce the salt and sugar in your diet.
- Eating too much sugar is a risk factor for cardiovascular disease. You should get about 50 per cent of your energy (kilocalories) from carbohydrates, but only 5 per cent should come from free sugars.
- Swapping sugary soft drinks for water or milk, and sometimes sugar-free soft drinks, is a good way of reducing sugar consumption.
- Some breakfast cereals are high in sugar.

> **Exam tip**
>
> You will need to understand the main risk factors that can cause obesity and cardiovascular disease. You should focus in your answers on the Eatwell Guide and the eight tips for eating well.

Now test yourself

TESTED ☐

1 Why are more school children becoming overweight? [1 mark]
2 What is cardiovascular disease? [2 marks]
3 State three ways in which diet may help to prevent cardiovascular disease. [3 marks]
4 List three ways in which sugar consumption may be reduced. [3 marks]

> **Typical mistake**
>
> Students sometimes focus too much on the medical conditions of obesity, CHD and stroke rather than how they may be prevented or treated by a healthy diet and lifestyle.

Bone health: rickets and osteoporosis

- To have healthy bones you need sunlight, exercise and a good diet, including foods that supply calcium and vitamin D to increase bone strength.
- Sunlight is absorbed by the skin to produce vitamin D.
- Diet has an important influence on a person's bone strength – from the very start of their development in the womb through to their early twenties.
- A deficiency of vitamin D and/or calcium causes rickets in children and osteoporosis in adults.

Rickets

- Children with rickets have weak and soft bones. It may cause the bones to change shape and bow outwards.
- There has been an increase in rickets in recent years; this may be due to children playing outdoors less.
- Low-fat milks contain less vitamin D than whole milk. Therefore whole milk is a better source of calcium and vitamin D.
- Children aged 6 months to 5 years should be given vitamin drops that contain vitamin D as it is difficult to get enough of this vitamin in the diet.

Osteoporosis

- Osteoporosis can be caused by a lack of calcium and vitamin D from early childhood through to the late twenties.
- Adults who develop osteoporosis are more likely to break a bone if they fall as their bones are weaker and more brittle.
- Exercise helps to prevent osteoporosis.

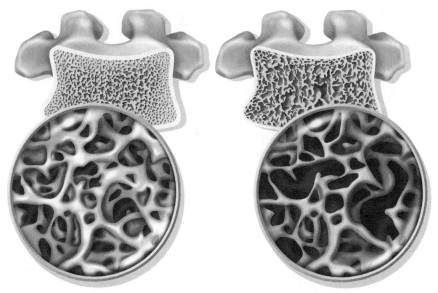

Figure 3.13 Osteoporosis makes bones fragile (right) and more likely to break

Dental health

- To keep teeth and gums healthy, you should have regular check-ups with your dentist.
- Tooth brushing should begin as soon as the first baby tooth appears.
- Children should have help brushing their teeth up until age 7.
- Teeth should be cleaned twice a day for about 2 minutes using a fluoride toothpaste.
- Eating foods high in sugar can cause tooth decay.
- If high-sugar foods are eaten regularly throughout the day as snacks, this can further increase the risk of tooth decay.
- Fruit juices should be limited to 150 ml per day; these can be diluted with water to reduce the overall acidity and sugar content per glass.

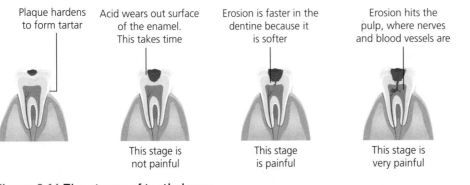

Plaque hardens to form tartar

Acid wears out surface of the enamel. This takes time

Erosion is faster in the dentine because it is softer

Erosion hits the pulp, where nerves and blood vessels are

This stage is not painful

This stage is painful

This stage is very painful

Figure 3.14 The stages of tooth decay

Plaque contains bacteria and forms on the teeth and gums. Over time, this plaque interacts with sugars and makes acid. The acid attacks the tooth enamel and causes tooth decay.

Typical mistake

Students are sometimes too general in their responses. They may write about eating more vitamins and minerals, but to get higher marks you should name these nutrients. For example, 'Vitamin D helps the absorption of calcium, which builds strong bones and teeth.'

Exam tip

Questions related to bone health and dental health will focus on how any bone and dental problems may be prevented. Therefore the importance of calcium and vitamin D should be understood. The need to reduce sugar to prevent tooth decay should be supported with practical examples (e.g. replace sugary drinks with milk or sugar-free fruit drinks).

Now test yourself

1 Which vitamin and mineral increase bone strength? [2 marks]
2 Why is sunshine good for bone health? [1 mark]
3 What is the name of the deficiency disease caused by a lack of calcium and vitamin D in (a) children and (b) adults? [2 marks]
4 Why is regular snacking on foods containing free sugars bad for your teeth? [3 marks]

Iron deficiency anaemia

- Iron deficiency anaemia occurs when there are not enough red blood cells. Red blood cells transport oxygen around the body. When there are not enough red blood cells, the body becomes short of oxygen.
- Anaemia may cause pale skin, breathlessness, heart palpitations, tiredness, dizziness or fainting. To find out if you have this type of anaemia, you would need a blood test.

Iron deficiency anaemia may be caused by a lack of iron in the diet. Girls and women who have heavy periods (menstruation) are at more risk of this type of anaemia.

- It may be treated with iron tablets or a diet high in iron and vitamin C. Vitamin C aids the absorption of iron and so should be included in dishes and meals containing iron (e.g. serve a side salad with red meat dishes).
- If it is left untreated, you are more likely to get infections as the body needs iron for the efficient functioning of its immune system.

Type 2 diabetes

REVISED

- Type 2 diabetes is the most common type of diabetes in the UK. Many people have type 2 diabetes and do not realise it. It causes the sugar in the blood to get too high.

The main symptoms of type 2 diabetes are:
- feeling tired all the time
- feeling thirsty
- passing more urine than usual.

You are more likely to develop type 2 diabetes if:
- you are overweight or obese
- you are over 40 years old
- you eat fatty, salty and sugary foods often
- you have high blood pressure
- you do not exercise regularly.

How to treat type 2 diabetes

- Follow the Eatwell Guide.
- Choose starchy carbohydrates as these will release sugar more slowly into the bloodstream.
- Have regular meals.

Typical mistake

Many students fail to mention the importance of vitamin C when writing about iron deficiency anaemia, as it is needed to absorb iron. For type 2 diabetes some students only write about reducing or removing free sugars from the diet; this is not accurate as all the current dietary guidelines are important in the control of type 2 diabetes.

Exam tip

Be able to plan meals for iron deficiency anaemia that are high in iron and vitamin C. Be able to plan meals for type 2 diabetes that include starchy, high-fibre foods, and are low in sugar and fat.

Now test yourself

TESTED

1 Why do people with iron deficiency anaemia get tired more easily? [1 mark]
2 Suggest a colourful side dish you could serve with a chicken burger to ensure the iron is absorbed more easily. [2 marks]
3 Why do teenage girls need more iron than teenage boys? [1 mark]
4 Name three risk factors for developing type 2 diabetes. [3 marks]

4 Cooking of food and heat transfer

Why food is cooked

- Cooking food makes it **safe** to eat because it destroys harmful bacteria.
- Cooking food improves the **flavour**, **appearance** and **smell** of food.
- Some foods are easier to **chew** and **digest** when they are cooked.
- Cooking food extends its **shelf life** meaning it can be stored for longer.
- Cooking food using different cooking methods **adds a variety** of flavours and textures to the diet.

The ways that preparation and cooking affect food

REVISED

Heat can change the appearance, colour, flavour, texture and smell of food.

Baking bread

How baking affects bread dough:
- **Appearance** – bread dough will rise.
- **Size** – bread dough will increase in size.
- **Colour** – bread dough will turn golden brown.
- **Texture** – bread dough will become light and springy. The crust will become crispy.
- **Smell** – baking bread dough will produce an appealing aroma.

Figure 4.1 **A loaf of bread**

Roasting a chicken

How roasting affects chicken:
- **Texture** – meat will change from soft to firm.
- **Smell** – cooking meat will produce an appealing aroma.
- **Colour** – chicken will turn from pink to white and brown.
- **Size** – meat will shrink.
- **Appearance** – the surface will be crispy and clear juices will be released.

Figure 4.2 **A roast chicken**

How heat is transferred to food

- Food is cooked by heat energy.
- The three ways that heat energy can be passed through food are **conduction**, **convection** and **radiation**.
- Most food is cooked by a combination of the methods of heat transfer.

Conduction

- **Conduction** is when the heat travels through solid materials (e.g. metals and food).
- Heat is conducted from molecule to molecule in a liquid or solid.
- The heat is transferred into the food by contact with a heat source or something very hot.
- Metals are excellent conductors of heat. Cooking pans, utensils and baking trays are made from steel, copper or aluminium.
- Water is a good conductor of heat. Food placed in boiling water will cook quickly.
- Plastic and wood are poor conductors of heat. Pan handles or tools for stirring food that is cooking are made from plastic or wood because they remain cool during heating.

Convection

REVISED

- **Convection** is when heat travels through air or water.
- A **convection current** is the movement of heat in water or in the air.
- Convection currents happen because hot air rises and cool air falls.
- Ovens are heated by convection currents.
- The hottest part of the oven is the top shelf and the coolest part is the bottom shelf, except for fan ovens where all the shelves are about the same temperature.
- Baking, boiling, poaching and steaming all use convection currents to transfer heat into food.

Radiation

REVISED

- **Radiation** is when heat rays directly warm and cook food.
- Heat travels from one place to another.
- Food that is grilled or toasted is cooked by radiation.
- Radiation requires a direct heat source (e.g. a grill or barbecue).
- Cooking food in a microwave oven uses radiation to cook the food.

Exam tip

You may be asked to discuss the advantages or disadvantages of microwaves. Try to give a balanced response to this question. Remember to:
- include both the advantages and disadvantages
- write in sentences
- look at the marks awarded to help you decide how many points to make.

Revision activity

Write down some typical meals you have eaten (e.g. pizza, lasagne, curry).

Identify and explain the different methods of heat transfer used to prepare the meals.

Now test yourself

TESTED

1 List four reasons why we cook food.   [4 marks]
2 State the three methods of heat transfer.   [3 marks]
3 Describe three ways in which cooking affects the sensory qualities of food.   [3 marks]
4 Describe how food is cooked by convection currents.   [2 marks]

Typical mistake

In questions referring to the methods of heat transfer, candidates tend to provide the method of cooking, such as 'boiling' or 'steaming' rather than the specific terms 'convection' and 'conduction'.

Choice of cooking methods

- The ways in which we cook food can be divided into the following groups: **cooking with water**, **cooking with 'dry' heat** and **cooking with fat**.

Table 4.1 How heat is transferred during cooking

Way of cooking	Heat transfer	Examples
Cooking with water	Heat will pass through water very quickly It is transferred by conduction and convection currents	The cooking methods that use water are blanching, boiling, braising, poaching, simmering and steaming
Cooking with 'dry' heat	Heat will pass through the air in convection currents in the oven and as radiation from a grill or barbecue	The cooking methods that use 'dry' heat are baking, barbecuing, char-grilling, dry-frying, roasting and grilling
Cooking with fat	Heat will pass through oil or fat by conduction or convection currents	The cooking methods that use fat are shallow frying and stir-frying

The cooking methods

REVISED

Table 4.2 The cooking methods

Cooking method	Definition	Examples of food cooked by this method
Baking	Cooking food by dry heat in an oven	Pastries, breads and doughs
Barbecuing or char-grilling	Cooking food using glowing-hot charcoal	Meats, vegetables
Blanching	Cooking quickly in boiling water and then cooling immediately	Vegetables, meats
Boiling	Cooking food in boiling water	Vegetables, pasta, rice, eggs
Braising	Cooking food in the oven in a covered container with a liquid that forms a sauce	Meats, vegetables
Dry frying	Frying food in a frying pan without fat or oil	Sausages, bacon, sesame seeds, nuts
Grilling	Cooking food by applying heat to its surface	Bread, kebabs
Poaching	Cooking food in a hot liquid, just below boiling point	Eggs, fish
Roasting	Hot air and a small amount of fat cook the meat or vegetables in the oven	Potatoes, meats, vegetables
Shallow frying	Cooking food in a small amount of fat or oil in a shallow pan	Meats, vegetables, eggs
Simmering	Cooking delicate foods (e.g. eggs, fish, fruit) gently in water just below boiling point	Vegetables
Steaming	Using the steam from boiling water to cook food	Fish, vegetables
Stir-frying	Cooking food in a wok or frying pan with a small amount of oil	Vegetables

Exam practice answers and quick quizzes at **www.hoddereducation.co.uk/myrevisionnotes**

The advantages and disadvantages of cooking with water

- Cooking with water involves relatively low temperatures.
- Water boils at 100°C.

Table 4.3 Advantages and disadvantages of cooking with water

Cooking method	Advantages	Disadvantages
Blanching	Very quick so the loss of nutrients is small Keeps the crisp texture and colour of vegetables	Vitamin C, vitamins from the B group, iron and calcium will leach into the cooking water
Boiling	Quick so the loss of nutrients is small	Vitamin C, vitamins from the B group, iron and calcium will leach into the cooking water Overcooking will turn vegetables mushy Flavours and colour will be lost
Braising	Softens and tenderises meats Flavours are enhanced Cooking liquid is eaten with the dish so water-soluble nutrients will be eaten	Colour can be lost Vitamin C, vitamins from the B group, iron and calcium will leach into the cooking water Very slow cooking method so the loss of nutrients is large
Poaching	Softens and tenderises food Flavour is retained	Some colour will be lost Vitamin C, vitamins from the B group, iron and calcium will leach into the cooking water Requires skill as it is easy to overcook food
Simmering	Softens food Tenderises meat, giving a more appetising texture	Flavour will leach into the water Vitamin C, vitamins from the B group, iron and calcium will leach into the cooking water Takes a long time so nutrient losses are great
Steaming	No contact with water so vitamin C, vitamins from the B group, iron and calcium will not leach Gives a light and delicate texture to food Tenderises meat	Fruit and vegetables may lose their colour Requires skill as it is easy to overcook food Some loss of water-soluble vitamins from the heat

The advantages and disadvantages of cooking with dry heat

● The oven, hob or grill can be used to produce dry heat.

Table 4.4 Advantages and disadvantages of cooking with dry heat

Cooking method	Advantages	Disadvantages
Baking	Does not affect calcium and iron Tenderises meat, giving it a softer, more appetising texture Gives a crispy texture and a golden-brown colour to the surface of food	Vitamin C and vitamin B1 are lost due to the heat
Barbecuing or char-grilling	Food becomes crispy and the flavour and smell are improved	Fat melts and runs out of the food so some vitamin A and D will be lost Not suitable for tough or very thick cuts of meat Overcooking can produce a bitter flavour and a black colour
Dry frying	Vitamins A and D are retained Flavour and smell are improved Food will crisp and brown No added fat, so healthier than other methods of frying	Vitamin C and vitamin B1 are lost due to the heat
Grilling	Most vitamin C, vitamins from the B group, iron and calcium will be retained Fat drains off the food Food will be crisp and brown	Some vitamin C and vitamin B1 are lost due to the intense heat Fat-soluble vitamins are lost when the fat melts because they will run out of the food Not suitable for tough cuts of meat or very thick foods Overcooking produces a bitter flavour and a black colour
Roasting	Does not affect calcium and iron Surface of meat will be brown and crisp Tenderises meat, giving it a softer texture	Some vitamin C and vitamin B1 are lost due to the heat and long cooking time Fruit and vegetables may lose their colour Adds fat to food

The advantages and disadvantages of cooking with fat

● Low-fat spreads are not suitable for frying because they contain water.
● Vegetable oils, **ghee**, butter and lard can all be used.

Table 4.5 Advantages and disadvantages of cooking with fat

Cooking method	Advantages	Disadvantages
Roasting	Does not affect calcium and iron Surface of meat will brown and crisp Tenderises meat, giving it a softer texture	Some vitamin C and vitamin B1 are lost due to the heat and long cooking time Fruit and vegetables may lose their colour Adds fat to food
Shallow frying	Most vitamins and minerals are retained Produces an interesting crispy texture	Bright colours are reduced Fat content of the food will increase
Stir-frying	Quick, so vitamins and minerals are retained Crisp texture and good colour in vegetables is retained Small amount of oil means that it is healthy too	Overcooking can reduce the colour and make food too soft

Revision activity

The diagrams in Figure 4.3 show eight cooking methods: steaming, boiling, simmering, braising, poaching, baking, shallow frying and barbecuing.

Copy the diagrams, label them and describe how heat is transferred in each example. Rate the method for vitamin retention using the terms 'poor', 'average', 'good' or 'excellent'.

Figure 4.3 Heat transfer

Now test yourself   TESTED ☐

1 Explain how heat travels when:
   (a) cooking with water   [2 marks]
   (b) cooking with 'dry' heat by conduction, when dry frying food, e.g. bacon in a frying pan   [2 marks]
   (c) cooking with fat   [2 marks]
2 Describe three advantages and three disadvantages of cooking with water.   [6 marks]
3 Describe three advantages and three disadvantages of cooking with dry heat.   [6 marks]
4 Describe three advantages and three disadvantages of cooking with fat.   [6 marks]

Typical mistake

Although vitamin C and the vitamin B group are water soluble it is their sensitivity to heat that is most important. Wherever possible, suggest that fruit and vegetables should be eaten raw or with minimal cooking for maximum vitamin retention.

5 Functional and chemical properties of food

Proteins

Protein denaturation

- **Denaturation** is a change in the structure of a protein.
- Protein is denatured by the action of cooking (using heat), adding an acid or whisking. These actions change the shape and structure of proteins.
- Denaturation is partially reversible. For example, when egg white is whisked, it incorporates air to form a foam. When the foam is left to stand, it can collapse back to form liquid egg white again.
- Some types of denaturation are irreversible – for example, heating and adding acid to foods make changes to the protein that are permanent.

Using an acid

- **Marinating** is the process of soaking meat or vegetables in liquid before cooking, giving food **flavour**, keeping it **moist** and making it **tender**.
- The marinating liquid can be **acidic** or **salty**. **Lemon juice** and **vinegar** are acidic marinades.
- The acid causes tough meat fibres to break down and moisture to be absorbed into the meat, making the meat juicy and tender.
- Acids can also be used to denature the protein in milk. This is known as **curdling**. Curdled milk develops a slightly lumpy appearance because all the protein in the milk has clumped together. Curdling will occur naturally when milk starts to 'go off', but it can also be caused by adding an acid, heat or enzymes to milk.
- Curdling is used in the cheese-making process to separate the protein 'curds' away from the liquid 'whey' part of the milk. The curds are then pressed into cheese.

Using mechanical action

- The physical act of **whisking** will cause a protein to denature.
- Eggs are a good source of protein and will stretch when whisked.
- The protein stretches into strands and forms a structure that allows air to be captured. Tiny air bubbles are held together by a mesh of protein. With continued whisking, a **foam** is produced. This is a temporary change in the structure of eggs.
- The physical act of **kneading** in the bread-making process can also cause proteins to denature. The protein in the flour becomes very stretchy when the bread dough is kneaded. This is a permanent change in the structure of the dough.

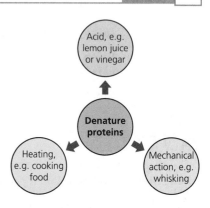

Figure 5.1 Ways of denaturing proteins

Using heat

- Proteins will also denature if **heat** is applied to them.
- This process is known as the **coagulation** of protein – for example, when egg white is cooked it changes from a liquid, becomes firmer and sets.

Protein coagulation

REVISED

- **Coagulation** is when the protein in food sets. This occurs when the protein is heated. If protein is heated too much, it will become hard, tough and difficult to digest.
- Heated protein foods will coagulate in different ways.

Table 5.1 The effect of heat on different protein foods

Protein food	Effect of heat
Meat, poultry and fish	Meat proteins shrink
	Overheating makes meat chewy
	Slowly heating meat proteins in a liquid changes collagen to gelatine, e.g. braising
Eggs	Egg white becomes solid and turns white at 60°C
	Egg yolk becomes solid and dry at 70°C
Wheat	Gluten sets, giving cakes, breads and biscuits their firm structure
Milk	Milk proteins will set and form a skin on the milk just below boiling point
Cheese	Cheese protein becomes rubbery if overheated

Figure 5.2 Egg coagulation

Gluten formation

REVISED

- The proteins **gliadin** and **glutenin** are found in wheat flour. They form **gluten** when mixed with water and kneaded.
- Bread-making flour, sometimes called **strong flour**, contains a large amount of both these proteins.
- These proteins give bread dough special qualities:
  - ○ glutenin gives the dough strength and elasticity
  - ○ gliadin binds the dough together into a sticky mass.

> **Typical mistake**
>
> Remember the importance of using **strong flour** to make bread.
>
> Plain flour or self-raising flour will not produce good results because they are low in protein.

Foam formation

REVISED

- Eggs are excellent at **foam** formation. You can whisk egg whites, egg yolks or whole eggs into a foam.
- A foam is when a gas is spread throughout a liquid, and whisking eggs produces a **gas-in-liquid** foam.
- Whisking makes the protein in the egg white unravel and **denature**. This allows tiny bubbles of air to be incorporated into the egg white, making an egg-white foam.
- If whisked for long enough the egg white will stand in soft peaks. When you heat the foam, the tiny air bubbles expand and the egg protein **coagulates** (sets) around them. This gives firmness to the foam. Examples of foods that rely on this property are meringues and soufflés.

Figure 5.3 Mechanical action produces a foam

Now test yourself

TESTED ☐

1 List four properties of proteins. [4 marks]
2 Give two reasons why marinades are used. [2 marks]
3 Describe how heat affects eggs. [4 marks]
4 What is syneresis? [2 marks]
5 Name the two proteins found in strong flour. [2 marks]

> **Exam tip**
>
> For the highest marks the correct use of the subject-specific terms is essential.
>
> The four key terms are **denaturation**, **coagulation**, **gluten formation** and **foam formation**.

Carbohydrates

Gelatinisation

REVISED ☐

- Some carbohydrates are starches (e.g. flour).
- When flour is mixed with a liquid and heated, as in a sauce, the mixture will thicken. This process is called **gelatinisation**.
- Gelatinisation happens for the following reasons.
  - ○ Starch grains do not dissolve in liquid. The liquid is heated to 60°C when gelatinisation begins and, by 80°C, most gelatinisation has taken place. Gelatinisation is not complete until boiling point (100°C) is reached.
  - ○ In the heated liquid the starch grains swell and break open.
  - ○ The liquid thickens and will set.
- Other starches that will thicken liquids are cornflour, arrowroot, tiny pasta shapes and potatoes.
- Two molecules in starch called amylose and amylopectin help it to thicken the liquid. When the starch granules break open, they release amylose. They bond with one another to form a gel when heated.

> **Exam tip**
>
> Always mention the temperatures for high marks. Gelatinisation happens between 60°C and 80°C, however it is not complete until 100°C (boiling point).

Table 5.2 What can affect gelatinisation?

Amount of liquid	A small amount of starch in proportion to the liquid will produce a runny sauce
	A large amount of starch in proportion to the liquid will produce a thicker sauce
Type of starch used	Cornflour is excellent for thickening because it is 100% starch
	Plain flour is good because it is about 75% starch
	Arrowroot produces a clear gel or glaze; it is gluten free but cannot be reheated
Temperature	Hot, moist conditions are needed. Starch grains cannot thicken cool liquids
	Starch grains need to be softened by heating
	Starch grains will start to thicken at 60°C
	Thickening stops at boiling point
	Sauces should be gently stirred and boiled for 2 minutes to complete gelatinisation
Stirring	Essential for a smooth, gelatinised sauce
	No stirring means that starch grains sink to the bottom of the pan and absorb the liquid round them
	No stirring produces a lumpy texture
Sugar	Sugar makes sauces runny
	Sugar stops the gel forming
Acids	Lemon juice should always be added to the sauce after it has boiled because it will break down the gel

Figure 5.4 Custard sauce produced by gelatinisation

Exam practice answers and quick quizzes at **www.hoddereducation.co.uk/myrevisionnotes**

Dextrinisation

REVISED

- **Dextrinisation** is when dry heat turns a starch brown. It occurs when starch is broken down into **dextrin** by **dry heat**.
- Dextrin adds a **sweet taste** to baked products.
- Dextrinisation happens in baking, grilling and toasting.
- Dextrinisation contributes to the colour and flavour of toast, bread and croissants.

Caramelisation

REVISED

Caramelisation occurs when heat is applied to sugar.
- Caramelised sugar is used to make fudge, toffee, jam and honeycomb.
- The surface of biscuits, breads and pastries is browned by caramelisation.
- Fruit, vegetables and meat contain sugar that will caramelise.
- Caramelisation occurs during dry heating, shallow frying, and the roasting of meat and vegetables.
- Caramelisation changes the colour and flavour of food.
- Caramelisation begins when:
  ○ sugar is heated and starts to melt; the water evaporates as steam
  ○ at 180°C the sugar turns from clear to dark amber
  ○ an attractive flavour and golden colour develop.
- Overheating produces a bitter taste and a burnt appearance. Eventually, pure carbon is produced.

Figure 5.5 Crème brûlée is a dessert with a hard sugar topping, achieved by using a blowtorch or a very hot grill to caramelise the sugar

Revision activity

Draw and label a series of simple diagrams to show the process of gelatinisation.

Use the words heat (60°C to 80°C), water, starch grains, swell, burst, absorb, thicken.

Typical mistake

Gelatinisation is the process of using starch to thicken a liquid.

Gelatin is a protein made from animal bones that will set a jelly.

Exam tip

Candidates who transfer knowledge from aspects of their practical work to the exam can achieve more marks. Always think about the practical work you have completed to help you remember the processes of gelatinisation, dextrinisation and caramelisation.

Here are some examples:
- gelatinisation – making a white sauce for a pasta dish in a saucepan
- dextrinisation – baking some biscuits or scones in the oven
- caramelisation – browning some onions in a frying pan.

Now test yourself

TESTED

1 Name the temperatures at which gelatinisation begins and ends.   [2 marks]
2 Describe three factors that will affect gelatinisation.   [6 marks]
3 Name the two starch molecules that cause gelatinisation.   [2 marks]
4 Explain how dextrinisation improves the flavour and colour of food.   [2 marks]
5 Describe what happens when moist or dry heat is applied to sugars.   [2 marks]

Fats and oils

Fats and oils have four important functions in food preparation and cooking:

Exam tip

Use the acronym 'PEAS' to remember the four properties of fats and oils:
P plasticity
E emulsification
A aeration
S shortening.

Shortening

REVISED

- Fats and oils give biscuits, shortbread and pastries a **crumbly** texture.
- The ability of fat to do this is called **shortening**.
- The best fats for shortening are butter, lard and baking margarine.
- The fat or oil:
  ○ gives the flour particles a waterproof coating
  ○ prevents the flour from absorbing water and reduces the development of gluten.

Aeration

REVISED

- **Aeration** is when air is trapped in a mixture.
- Air needs to be added to mixtures to give a **springy texture**.
- In cake making:
  ○ fat and sugar are **creamed** together using a whisk or wooden spoon
  ○ bubbles of air are enclosed in the mixture; this makes a stable **foam**
  ○ baking gives the cake a springy texture.

Plasticity

REVISED

- Fats melt and soften at different temperatures; they have different melting points.
- The **plasticity** of fat is linked to their different melting points.
- Plasticity affects spreading, creaming and shortening.
- Fats selected for shortening must have good plasticity. Good plasticity means that the fat will spread over a large area of flour and coat it with a film of oil.
- Fats can be processed to alter their melting point (e.g. fridge-spreadable fats).

Emulsification

REVISED

- Fats and oils do not mix with water.
- Fats and oils are '**immiscible**', which means they cannot be mixed.
- Forcing two immiscible liquids together is called **emulsification**.
- An **emulsion** is a special type of liquid that allows tiny droplets of one liquid, such as oil, to be spread throughout another liquid, such as water.

Table 5.3 Types of emulsion

Types of emulsion	Description
Oil-in-water emulsion	This forms when the amount of **water** is **more than** the amount of **oil**
	Tiny droplets of oil are spread throughout the water
	Milk is a good example of this type of emulsion; it contains about 4% fat and 96% water
Water-in-oil emulsion	This forms when the amount of **oil or fat** is **more than** the amount of **water**
	Tiny droplets of water are spread through the fat or oil
	Butter and fat spreads are water-in-oil emulsions

Emulsifiers

In all foods, we want emulsions that will last and not separate out. Sometimes, if liquids are allowed to stand for a long time, the oil or water will separate out from the mixture. This can be seen as two layers.

- The formation of a stable emulsion will depend on the presence of an **emulsifier**.
- An emulsifier is a substance that will allow two immiscible liquids (substances that do not mix) to be held together.
- An emulsifier has two parts. One part of the emulsifier attracts the water and one part attracts the oil. This combination holds the oil and water together, so that they can combine to form a **stable emulsion**.
- Emulsions are important in ice cream, mayonnaise and salad dressing.
- **Lecithin**, found naturally in egg yolk, is an emulsifier. Lecithin is used to produce mayonnaise. It helps to stop the oil and vinegar from separating out in the mayonnaise.

Now test yourself

TESTED ☐

1 Explain why shortening is important in baked products. [2 marks]
2 Describe how air is added to a cake mixture when using the creaming method. [2 marks]
3 List three ways that plasticity affects fats. [3 marks]
4 Describe the two different types of emulsion. [4 marks]
5 Explain how emulsifiers work. [4 marks]

Fruit and vegetables

Enzymic browning

- Enzymes cause changes in food.
- There are many different types of enzyme, but they are usually proteins.
- If the shape of the enzyme changes it will no longer work. We say the enzyme has been **denatured**.

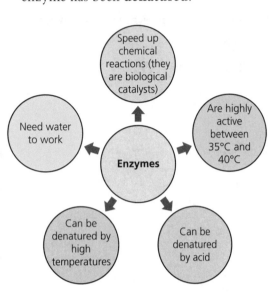

Enzymes

- Speed up chemical reactions (they are biological catalysts)
- Are highly active between 35°C and 40°C
- Can be denatured by acid
- Can be denatured by high temperatures
- Need water to work

Figure 5.6 Enzymes: key facts

> **Typical mistake**
>
> A common error in exams is to write that enzymes are killed at high temperatures. Although enzymes are made by living things, they are not living things and **cannot be killed**.

- **Enzymic browning** is an important reaction that occurs in fruit and vegetables. Some fruits and vegetables will turn **brown** because of the action of enzymes.
- Enzymic browning damages the colour, taste, flavour and nutritional value of fruit and vegetables.
- **Over-ripening**, **cutting**, **bruising** and **slicing** fruit and vegetables will lead to enzymic browning.
- **Blanching** is used to stop enzymic browning. It is the heating of fruit or vegetables in boiling water for a short time to destroy the enzymes before plunging them into iced water to stop the cooking process.

Table 5.4 Methods of slowing down enzymic browning

Refrigeration and chilling	At temperatures below 7°C, enzyme activity slows down
Freezing	Freezing slows down but does not stop enzyme activity
	After thawing, the enzyme activity will restart
Change pH	Lowering of the pH by adding acid stops the enzyme activity
	Enzymes are denatured
	Acids include lemon juice or vinegar
Dehydration	Removing water from the product stops the enzyme activity
Blanching	Boiling water denatures the enzymes

Figure 5.7 Enzymic browning of apples

Oxidation

- **Oxidation** is the loss of water-soluble vitamins on exposure to heat/air.
- **Enzymes** can cause foods to spoil by the process of oxidation.
- Oxidation occurs when food comes in to contact with air/heat.
- Peeling, slicing, shredding, chopping and grating will damage cell walls and expose enzymes to the air, causing oxidation.
- Using a blunt knife will damage more cell walls when chopping fruit and vegetables by exposing more enzymes to the air and causing oxidation.
- Using reduced oxygen or a **modified atmosphere** for storage (e.g. a salad bag) will slow down oxidation.
- Oxygen and the enzymes in fruit and vegetables cause loss of the water-soluble vitamins (B group and vitamin C).

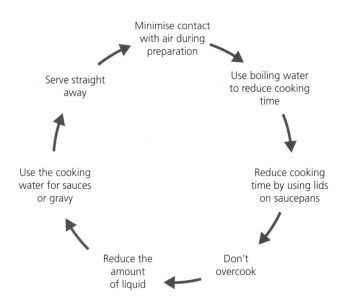

Figure 5.8 Reducing oxidation and loss of water-soluble vitamins

Now test yourself

TESTED

1 Suggest three ways that enzymic browning can occur in fruit and vegetables. [3 marks]
2 Describe two ways in which enzymic browning can be slowed down. [4 marks]
3 Which vitamins are lost due to oxidation? [1 mark]
4 Describe two methods that will help to reduce oxidation in fruits and vegetables. [4 marks]

Raising agents

- **Raising agents** are added to mixtures to make them rise.
- Raising agents are used in cakes, pastries, breads and sometimes biscuits to give a soft, springy texture.
- Raising agents work by introducing **gas** into a mixture.
- When you heat a mixture; the gas expands and makes the mixture rise.
- When the mixture cools and sets, the baked item has a light, open texture.
- The gases used are air, steam and **carbon dioxide**.
- The three types of raising agent are **chemical**, **mechanical** and **biological**.

Chemical raising agents

REVISED

- **Chemical raising agents** produce carbon dioxide when they are heated with a liquid.
- The two most common chemical raising agents are **baking powder** and **bicarbonate of soda**.

Table 5.5 Chemical raising agents

Type	Raising agent	Uses
Baking powder	A mixture of two chemicals Reacts with moisture and heat Produces the gas carbon dioxide	**Self-raising flour** has baking powder added which, when heated, produces carbon dioxide
Bicarbonate of soda	Needs moisture, heat and an acid to work Produces carbon dioxide Has a soapy taste	Soda breads Gingerbread Fruit cake Chocolate cake Carrot cake

Mechanical raising agents

REVISED

- **Mechanical raising agents** use air and steam.

Air

Table 5.6 Mechanical raising agents

Method	Process of adding air	Examples
Whisking	High-speed whisking Traps air bubbles	Roulades Meringues Whisked sponges
Beating	Using a spoon Air bubbles are trapped in a liquid	Batter
Folding	Air is trapped between the layers during folding	Flaky pastry
Sieving	Traps air in flour	Cakes, pastries and scones
Creaming	Beating fat and sugar together traps tiny air bubbles into the mixture	Cakes
Rubbing in	Rubbing fat into flour	Pastry and scones

Steam

- Steam is produced during cooking from water or liquids as the mixture reaches boiling point.
- Steam is used in making Yorkshire puddings, and choux, puff and flaky pastry.

Figure 5.9 Yorkshire puddings rely on steam for their light texture

Biological raising agents

REVISED

- **Yeast** is a biological raising agent used in bread, doughnuts and buns.
- Yeast is a type of fungus.
- Yeast produces the gas carbon dioxide by **fermentation**.
- Fermentation is when carbon dioxide is produced, which will make the dough rise.
- Yeast needs warmth, food and liquid to ferment.

There are two main types of yeast: fresh and dried.
1 **Fresh yeast** is a firm, moist, cream-coloured block that must be stored in a refrigerator. It is blended with warm water when required for dough making.
2 **Dried yeast** takes the form of small granules of yeast, which will keep for many months.

There are two types of dried yeast:
1 **fast-acting 'easy blend'** is usually mixed with the flour during the dough-making process
2 **active dried yeast** needs to be mixed with warm water and sugar before use.

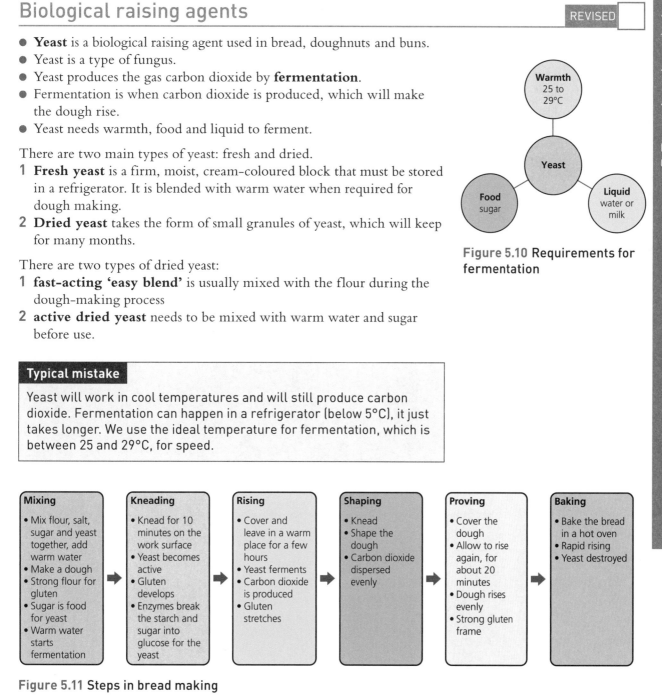

Figure 5.10 Requirements for fermentation

> **Typical mistake**
>
> Yeast will work in cool temperatures and will still produce carbon dioxide. Fermentation can happen in a refrigerator (below 5°C), it just takes longer. We use the ideal temperature for fermentation, which is between 25 and 29°C, for speed.

Mixing	Kneading	Rising	Shaping	Proving	Baking
• Mix flour, salt, sugar and yeast together, add warm water • Make a dough • Strong flour for gluten • Sugar is food for yeast • Warm water starts fermentation	• Knead for 10 minutes on the work surface • Yeast becomes active • Gluten develops • Enzymes break the starch and sugar into glucose for the yeast	• Cover and leave in a warm place for a few hours • Yeast ferments • Carbon dioxide is produced • Gluten stretches	• Knead • Shape the dough • Carbon dioxide dispersed evenly	• Cover the dough • Allow to rise again, for about 20 minutes • Dough rises evenly • Strong gluten frame	• Bake the bread in a hot oven • Rapid rising • Yeast destroyed

Figure 5.11 Steps in bread making

Exam tip

Be clear about the gases involved: steam, carbon dioxide and air.
● Steam is produced by boiling water or liquids.
● Carbon dioxide is produced by chemicals or yeast.
● Air is added during the making.

Now test yourself

TESTED ☐

1 Name three gases that raising agents produce. [3 marks]
2 Name two chemical raising agents. [2 marks]
3 Describe four ways that air can be added to a mixture. [4 marks]
4 Describe three conditions required for yeast to ferment. [6 marks]
5 Describe the two different types of yeast. [4 marks]

6 Food spoilage and contamination

Micro-organisms and enzymes

- **Micro-organisms** (micro means small) are tiny organisms that can spoil food.
- The micro-organisms are **yeasts**, **moulds** and **bacteria**.
- Enzymes can also spoil food, as well as allowing fruits and vegetables to ripen.
- Food spoilage happens when food is contaminated by yeasts, moulds or bacteria.
- Food spoilage may also be caused by enzymic activity.

Micro-organisms

REVISED

The micro-organisms – yeasts, moulds and bacteria – require the following conditions to be able to grow:

- **food**
- **moisture**
- **warmth**
- **time**.

Yeasts

- Yeasts are single-celled fungi that reproduce by '**budding**'.
- Budding means the yeast cell grows a bud that becomes bigger until it eventually breaks off and becomes a new yeast cell.
- Yeasts can grow with oxygen (aerobic yeasts) or without oxygen (anaerobic yeasts).
- Yeasts prefer moist, acidic foods (e.g. oranges).
- Yeasts can grow in high concentrations of sugar and salt (e.g. jam, honey and chutneys).
- Yeasts grow best in warm conditions (around 25–29°C).
- Yeasts can grow at refrigerator temperatures (0 to below 5°C).
- Yeasts are destroyed at temperatures above 55°C.

Figure 6.1 Yeasts reproduce by budding

Moulds

- Moulds are tiny fungi that produce thread-like filaments that help the mould to spread around food.
- Moulds can grow in foods with high sugar and salt concentrations.
- Moulds grow best between 20°C and 30°C.
- Moulds can grow at refrigerator temperatures (0 to below 5°C).
- Mould growth can be speeded up by high humidity and fluctuating temperatures.
- Moulds can grow on fairly dry food, such as hard cheeses (e.g. Cheddar cheese).
- Moulds often spoil foods such as bread and other bakery products.
- Do not eat foods spoiled with moulds – even if scraped off, the mould may have spread harmful substances deep into the food.

Bacteria

Bacteria are single-celled micro-organisms. They can be divided into three groups:

1 harmless bacteria
2 pathogenic bacteria
3 food spoilage bacteria.

- Bacteria can be found everywhere around you: on your skin, in food, in soil, in water and in the air.
- Most bacteria are harmless, but some are **pathogenic** (harmful) and can cause food poisoning.
- Other types of bacteria cause food to decay; these are called **food spoilage bacteria**. They cause food to smell and lose its texture and flavour.
- Bacteria grow best with warmth (around body temperature, 37°C). They prefer moist conditions on neutral foods containing protein.

Bacteria reproduce by a process called '**binary fission**' in which one bacterium divides into two bacterial cells.

Bacterial growth can be slowed down or prevented by:

- storing food in the refrigerator
- cooling cooked perishable food quickly (within 90 minutes) before refrigerating
- reheating leftover food to 75°C, once only
- using high concentrations of salt (e.g. for bacon)
- using high concentrations of sugar (e.g. for jam)
- using high concentrations of acid (e.g. for pickled eggs).

Which foods do bacteria often spoil?

Foods that provide the bacteria with moisture, a high protein content and a neutral pH provide ideal conditions for bacteria growth, especially in a warm environment. Foods that provide these conditions and are ready to eat without further heating or cooking are known as **high-risk foods**.
For example:

- cooked meat and poultry
- cooked meat products such as gravy, soup and stock
- milk and eggs and dishes made from them (e.g. unpasteurised soft cheese)
- eggs and dishes made from them (e.g. homemade mayonnaise)
- shellfish such as mussels, crabs and lobsters
- cooked rice.

Figure 6.2 Mould growing

Figure 6.3 Pathogenic bacteria (listeria) under a powerful microscope

Enzymes

- **Enzymes** are biological catalysts, usually made from proteins.
- Enzymes speed up chemical reactions.
- Enzymes can cause browning in foods that are bruised or cut open; this is called **enzymic browning**.
- Enzymes can be destroyed by heat or acids.
- **Blanching** means heating food in boiling water for a short time to destroy enzymes before plunging into iced water to stop the cooking process.
- Enzymes can still be active at freezer temperatures of -18°C or below.
- If foods such as apple slices or corn on the cob are not blanched before freezing they may spoil due to enzymic activity during storage in the freezer.
- Acids such as lemon juice may be used to prevent enzymic browning (e.g. on apple slices). The acid denatures the enzymes so they become inactive.

Oxidation

- Oxidation is the loss of water-soluble vitamins on exposure to heat/air.
- Enzymes can cause foods to spoil by the process of oxidation.

Exam tip

You need to be able to name the different micro-organisms along with their ideal growth conditions. Although some of these growth conditions are the same, there are some differences you should learn and understand the reasons why.

Typical mistake

You may be asked to state growth conditions for bacteria. A common mistake is to write 'heat' as one of the conditions. You would not gain a mark for this as heat is too vague and could include very hot temperatures above 63°C, which would destroy the bacteria. Instead write 'warmth' or 'around body temperature, which is 37°C'.

Now test yourself

1 List the three types of micro-organism. [3 marks]
2 Name two ways enzymes can affect food. [2 marks]
3 What are the four growth conditions needed by micro-organisms? [4 marks]
4 Explain why cooked chicken and traditional quiche are described as 'high-risk foods'. [4 marks]
5 Explain three different ways oxidation may be prevented when preparing and cooking fresh broccoli. [6 marks]

6 Food spoilage and contamination

The signs of food spoilage

Enzymic action

REVISED

(See also the section on enzymes on page 56.)

● Enzymes are found naturally in foods and, once bruised or cut open, enzymic browning occurs

● Enzymes can also cause over-ripening of fruits such as bananas.

Figure 6.5 Enzymic browning of avocados

Figure 6.6 Enzymes ripen bananas but can eventually spoil them

Yeast action

REVISED

(See also the section on yeasts on page 51.)

● Yeasts can grow on acidic, sweet foods (e.g. orange juice can ferment if it is not stored correctly)

● Yeasts may ferment honey that is not pasteurised.

Figure 6.7 Yeast may ferment orange juice if it is not stored correctly

Mould growth

REVISED

(See also the section on moulds on page 52.)

● Moulds like to grow on food in warm and moist conditions

● Mould grows easily on bread, cheese and soft fruits.

Figure 6.8 Moulds can grow on the surface of jam

Micro-organisms in food production

Micro-organisms are used to produce many different food products.

Yeast in food production

REVISED

(See also the section on yeast as a raising agent on pages 48–49.)

- Yeast is used as a raising agent in different foods, including bread, crumpets, doughnuts and current buns.
- Yeast is used in the brewery industry to ferment grapes into wine and beer into hops.

Figure 6.9 Yeast is used as a raising agent in this dough recipe

Mould in food production

REVISED

- Some cheeses are made with moulds to develop their colour and flavour (e.g. Brie, Camembert and Stilton cheeses).
- The use of moulds in food production provides a wider variety of flavours and textures to food products.
- Moulds are also used to ripen the surface of some sausages, which improves their smell and texture as well as extending their shelf life (e.g. salami).
- Moulds (along with yeast) are used in the production of soy sauce.

Figure 6.10 Some cheeses use moulds to provide distinct colours and flavours

Bacteria in food production

REVISED

How to make cheese

- Bacteria are used in cheese making in the form of a starter culture added at the beginning of the cheese-making process; along with rennet and salt.
- The starter culture of bacteria converts the milk sugar lactose into lactic acid. This begins the cheese-making process.

How to make yoghurt

- Harmless bacteria are added to sterilised, warmed milk (about 37°C).
- This is then kept warm to allow the bacteria to multiply over several hours.

Figure 6.11 Bacteria cultures are used to produce cheeses and yoghurt

Bacterial contamination

Sources of bacterial contamination

- Foods may become contaminated with bacteria from many different places.
- Many raw foods are sources of bacterial contamination.
- It is safest to assume all raw meat and their juices are contaminated with food poisoning (pathogenic) bacteria.
- Raw meat and their juices can contaminate other foods if they are allowed to come into contact with them directly or with equipment.
- Other types of raw food that carry bacteria are raw eggs, which may have bacteria on the inside and outside of their shells, as well as shellfish such as mussels and oysters.
- Plant foods such as rice and vegetables can be contaminated with bacteria from the soil.

Work surface and equipment contamination

- Work surfaces and equipment can become contaminated with bacteria from raw foods and unwashed hands.
- Bacteria are so small you cannot see them without a powerful microscope, so always clean work surfaces before you begin any food preparation.
- Colour-coded chopping boards and equipment help to prevent cross-contamination.
- Dishwashers can wash equipment at high temperatures, which destroys bacteria more efficiently than washing by hand.
- Dishcloths and tea towels should be washed in very hot water daily as they can spread bacteria around if they are not clean. Disposable cloths can be a better alternative.
- Anti-bacterial sprays can be used to sterilise work surfaces after they have been cleaned with hot soapy water.

Food handler contamination

When you are preparing and cooking food, you can contaminate the food with bacteria that are naturally present on you. To avoid this, you should:
- use tongs or other equipment to pick up food
- use a clean teaspoon and don't double dip when tasting food while cooking!
- not lick your fingers or eat while cooking; this will spread bacteria from your mouth to your hands, which can then contaminate food
- pick up utensils by their handles (e.g. don't put your hands inside cups to pick them up as this will contaminate them with bacteria).

Pest contamination

- Pests such as flies, insects, birds, mice and rats can contaminate food, storage and preparation areas. They carry bacteria on their bodies and in their urine and droppings.
- Pests can be kept out of kitchens by keeping windows closed and using extractor fans for ventilation.
- Don't leave dirty dishes lying around. Keep work surfaces and floors clean so there is no food source for pests. Don't leave food out overnight.
- Bins should be clean, have a lid and be emptied regularly.
- Domestic pets such as cats and dogs carry food poisoning bacteria and should not be allowed in food areas.

Figure 6.12 Rats carry pathogenic bacteria, which can lead to food poisoning

Waste food and rubbish contamination

- Bacteria can multiply in waste food and rubbish bins.
- Use bins with lids, use thick bags, empty bins regularly; bins should have a lid and be kept clean.

Food poisoning

REVISED

At-risk groups

For some groups of people, food poisoning can be very serious. These groups are:
- babies and young children
- pregnant women
- elderly people
- people with reduced immunity.

How is food poisoning caused?

Food poisoning is caused by:
- eating contaminated food or drinking contaminated water
- eating undercooked food, especially meat and poultry
- not keeping food chilled; perishable food should be kept in the fridge between 0°C and below 5°C
- not putting foods on the correct shelves in the fridge
- eating food that has been contaminated by someone with a septic cut or someone with food poisoning
- cross-contamination, which is when bacteria from raw food or unclean work surfaces come into contact with ready-to-eat food, making it unsafe to eat.

What are the symptoms of food poisoning?

The symptoms of food poisoning are:
- vomiting
- diarrhoea
- nausea
- stomach pains
- dehydration.

The symptoms may start within two hours of eating contaminated food or may begin many days afterwards.

The main types of food poisoning bacteria

Table 6.1 The main sources of food poisoning bacteria

Name of pathogenic bacteria	Foods bacteria is found in	Source of bacteria (where bacteria comes from)
Campylobacter	Poultry, milk and milk products	Unclean water Unpasteurised milk Bottled milk pecked by birds Raw poultry Sewage
E. coli	Undercooked meat, especially burgers and mince Unwashed contaminated fruit	Raw and undercooked meat, dirty water, sewage
Salmonella	Undercooked or contaminated cooked meat Beansprouts Unpasteurised milk Foods made from imported poultry and eggs	Intestines of ill people, animals, birds, raw meat, unpasteurised milk, imported poultry and eggs
Listeria	Pâté, cooked chicken, prepared salads, soft cheeses	Sewage, decaying vegetation and unclean water
Staphylococcus aureus	Unpasteurised milk, meat and meat products	Human food handlers touching, coughing or sneezing onto food, as well as human-to-human contact An infected person may contaminate ready-to-eat food

Typical mistake

You may be asked to name some 'high-risk' foods. A common exam mistake is to write about sources of food poisoning bacteria such as raw meat or poultry. High-risk foods are 'ready to eat' foods that can support the growth of food poisoning bacteria, and so correct answers would include cooked meat or poultry, homemade mayonnaise or cooked rice.

Exam tip

You will need to learn the names of the food poisoning bacteria along with the typical food sources (e.g. E. coli bacteria can be found in undercooked beef burgers).

Now test yourself

 TESTED

1 Give two examples of how enzymes can spoil food. [2 marks]
2 Apart from enzymes, name the three micro-organisms that can cause food spoilage. [3 marks]
3 State how each of the three micro-organisms are used in the production of named foods. [6 marks]
4 Name three sources of bacterial contamination. [3 marks]

7 Principles of food safety

Buying and storing food

Temperature control when storing food

REVISED

- The temperature control of food is very important.
- Many outbreaks of food poisoning are caused by high-risk food being kept for too long within the temperature danger zone. High-risk foods are the ready-to-eat foods that are most likely to cause food poisoning. They need to be stored in the fridge or freezer.
- The temperature **danger zone** is between **5°C and 63°C**. Bacteria can multiply between these temperatures. Keep high-risk foods out of the temperature danger zone.
- Below 5°C, bacteria may multiply very slowly or not at all. Keep **cold foods** cold **below 5°C** or in the **freezer at –18°C**.
- No bacteria multiply at temperatures above 63°C. Keep **hot foods** hot at **63°C or above**.
- Bacteria multiply the fastest at around 37°C (body temperature).
- Some foods, such as canned foods or breakfast cereals, are described as non-perishable and are normally stored in cupboards at around 17°C to 20°C. This is called **ambient** temperature or room temperature.
- **Fridge temperature** should be **0°C to below 5°C**.
- The safe temperature for the **reheating** and cooking of food is **75°C**.

Perishable foods

- Perishable foods include both raw and cooked foods with a fairly short shelf life; they usually need to be stored in the refrigerator or freezer (e.g. raw and cooked meat, poultry and fish, milk and eggs, bread, butter, low-fat spreads and yoghurts, cheeses, fruits, vegetables and salads).
- When shopping for perishable foods, keep them cool until you get home. The refrigerated items will be best stored in an insulated cold bag with ice packs.
- Some perishable foods have a 'use by' date (e.g. pre-packed meat, poultry, milk and yoghurts) and some have a 'best before' date (e.g. eggs, bread, butter and pre-packed vegetables).
- Non-perishable foods have a best before date, which indicates the recommended time in which the food should be eaten, but it is still safe to eat it after this date.

Figure 7.1 Prawn cocktail is a high-risk food and should be kept out of the temperature danger zone

Correct use of refrigerators (fridges)

REVISED

- Fridges keep food chilled, ideally between 0°C and below 5°C.
- At fridge temperatures micro-organisms grow very slowly or not at all as they do not have the warmth they need to multiply.
- High-risk foods should be stored in the fridge.
- The fridge temperature will vary from shelf to shelf, but all shelves should be below 5°C; you can place a thermometer on the different shelves to check this.
- Eggs have bacteria on their shells so should be stored in their box on the bottom shelf of the fridge.

- Place food in the fridge with spaces in between items so air can circulate and chill the food properly.
- Remember to follow date marks on foods and to rotate stock, so 'first in, first out'.
- Always cool food to room temperature before refrigerating.
- Open and close the fridge door as little as possible to avoid a rise in the fridge temperature.
- Wrap all food that is stored in the fridge to:
  - stop it drying out
  - prevent cross-contamination
  - prevent tainting by stronger foods (e.g. raw onion can taint other foods).

Types of refrigerator (fridges)

- Some fridges have an ice box; these will need to be defrosted to remove ice that has built up. This should be carried out about once a month following the manufacturer's instructions.
- Fridges without an ice box are called larder fridges and although these don't need to be defrosted, they should be emptied and cleaned about once a month.
- Bicarbonate of soda dissolved in hot water is good for cleaning fridges as it won't taint the food.

READY TO EAT FOODS
Such as dairy products, yoghurts, cream...

..cream cakes, butter/margarine, cooked meats, leftovers-covered, other packaged foods, e.g. coleslaw, tomato ketchup, jams etc.
TOP SHELVES AND MIDDLE SHELVES

RAW MEAT, POULTRY and FISH
Always cover and keep in sealed containers.
BOTTOM SHELVES

SALAD VEGETABLES, FRUIT & VEGETABLES
Keep ready to eat fruit and vegetables in sealed bags or containers. Always wash raw fruits and vegetables before use.
SALAD DRAWER

CORRECT FRIDGE TEMPERATURE 0°C TO BELOW 5°C

Figure 7.2 Where to store food in a fridge

Correct use of freezers

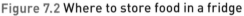
REVISED

- Freezers store foods at temperatures of -18°C or colder.
- Micro-organisms cannot multiply at these very cold temperatures (the bacteria become dormant), but most survive the freezing process.
- As the water is frozen, micro-organisms do not have any moisture to multiply.
- Enzymes may still be active at freezer temperatures, which is why some fruits (e.g. apple slices) and vegetables (e.g. corn on the cob) need to be blanched before freezing. See page 36 for more information on blanching.
- Once the frozen food is defrosted, micro-organisms (including food poisoning bacteria) come out of their dormant state and can start to multiply again.
- Foods may be stored in the freezer for up to one year, but follow date marks on shop-bought frozen food.
- A freezer needs to be defrosted about once a year to avoid a build-up of ice, which stops it working properly.
- Frost-free freezers should be switched off, emptied and cleaned once a year following the manufacturer's instructions.
- When freezing food at home:
  - use the fast-freeze button, which will speed up the freezing process as the temperature drops to as low as -25°C
  - wrap, label and date food to easily identify it, to prevent freezer burn and to ensure foods are used in date order
  - store food items with spaces in between to allow the cold air to circulate, and do not overload.

'Best before' and 'use by' dates

REVISED

- Most foods have a **best before** or a **use by** date mark to let the consumer know the freshness of the food.

'Best before' dates

- **Best before dates** relate to the sensory qualities of the food, such as flavour and texture, rather than food safety.
- It is considered safe to eat foods after the 'best before' date.
- Most foods with a 'best before' date are non-perishable, although eggs are the exception to this rule and should be eaten by the date printed on the box.
- Shops are allowed to sell foods after the best before date, with the exception of eggs.

'Use by' dates

- 'Use by' dates are found on foods that may be unsafe to eat after this date as they could cause food poisoning.
- Foods displaying 'use by' dates are usually perishable foods such as pre-packed raw and cooked meat, fish and poultry, milk, meat products, dairy foods and ready-prepared salads.
- It is against the law to sell food after its 'use by' date.

Figure 7.3 A food umbrella protects food from pests

Exam tip

Learn the key temperatures for the safe storage of food and the importance of keeping perishable foods out of the temperature danger zone.

Covering foods

REVISED

- All food stored in the fridge, freezer or cupboards should be wrapped or covered to keep it clean, avoid cross-contamination and to prevent it from drying out.
- Foil and cling film are suitable wrappings for covering foods, although cling film should not come into direct contact with high-fat foods to prevent the migration of plastics into the food, which may be harmful.
- When food is ready to be served, it may be covered to protect it from contaminants including pests, such as flies and wasps. Dust and other foods may also contaminate ready to eat foods if they are left uncovered.
- Food umbrellas are useful for keeping pests such as flying insects away from food.

Typical mistake

You may be asked to describe the differences between 'use by' dates and 'best before' dates. A common mistake is to write 'use by' dates are on perishable foods and 'best before' dates are on non-perishable foods, as you will gain limited marks. Instead think about the following points.

- A 'best before' date means that food is at its best quality before this date. With the exception of eggs, it is safe to be eaten after this date.
- A 'use by' date means that food must be consumed by this date to prevent food poisoning.
- Give examples of foods with a 'best before' date (e.g. bread, breakfast cereals, canned foods with a 'use by' date, e.g. sausage rolls, yoghurts, raw chicken).

Now test yourself

TESTED

1 Complete the missing temperatures:
   - A freezer should be _____°C or below
   - Food should be chilled (refrigerated) between _____ and _____°C
   - The temperature danger zone is between _____°C and _____°C
   - Food should be cooked to an internal temperature of _____°C
   - Food should be reheated to an internal temperature of _____°C   [7 marks]
2 Give three rules for the correct use of the refrigerator (fridge). [3 marks]
3 Give three rules for the correct use of the freezer.   [3 marks]
4 Explain why it is safe to eat foods after their 'best before' date. [3 marks]
5 Give two reasons why food should be covered before refrigeration.   [2 marks]

Preparing, cooking and serving food

When preparing, cooking and serving food, there are some basic rules that need to be followed to prevent food poisoning.

Personal hygiene

- Personal hygiene is very important for anyone preparing, cooking and serving food.
- High standards of personal hygiene will mean food poisoning is less likely to occur.
- The most common way bacteria are spread onto food is by your hands, so hand washing is very important.
- Hand washing should be thorough, making sure you wash your fingertips and thumbs as well as your wrists.
- Your hands should be dried with a disposable paper towel or hot air drier to reduce the spread of bacteria.
- Your fingernails should be kept short and clean. You should not wear nail varnish when preparing food as it may flake off into the food.

When should you wash your hands?

You should wash your hands:
- before you start any food preparation
- after touching your hair or face
- after using the toilet
- after using a handkerchief or tissue to cough or blow your nose
- after cleaning or putting rubbish into the bin
- after handling raw meat, poultry, vegetables or eggs
- after eating or drinking.

Figure 7.4 How to wash your hands properly

Personal hygiene rules for the kitchen

- Don't cough or sneeze near food.
- Don't touch your head, especially your mouth, nose and ears.
- Don't brush your hair in the kitchen or with your apron on.
- Long hair should be tied back or covered.
- Wounds such as cuts and scratches should be covered with a coloured waterproof plaster.
- Wear a clean apron (to protect the food from bacteria on your clothes).
- Don't prepare food if you are unwell with diarrhoea or coughs and colds, as you could spread bacteria onto food.

Clean work surfaces

- Work surfaces in the kitchen should be kept clean, to reduce bacteria to a safe level.
- Before you begin any food preparation, work surfaces should be cleaned with hot soapy water using a clean cloth.
- Anti-bacterial washing-up liquid will clean the work surfaces and destroy bacteria.
- Anti-bacterial sprays may be used, but these do not remove grease and dirt so should be used after cleaning with a detergent such as washing-up liquid.

Separating raw and cooked food

- Colour coding in the kitchen helps to prevent the spread of bacteria from raw to cooked food. See page 56 for more information on colour coding.
- All ready-to-eat foods should be kept away from raw foods that require cooking.
- Some raw foods (e.g. raw meat) normally contain lots of bacteria. Cooking destroys most of these bacteria.
- If cooked food comes into contact with some raw foods it can become contaminated; this is called **cross-contamination** and could lead to food poisoning.
- For example, when preparing, cooking and serving food at a barbecue you should have separate utensils and plates to avoid cross-contamination of bacteria from the raw to the cooked food.

Figure 7.5 Use separate utensils for raw and cooked food

Cooking food for the correct time

- Some food is safe to be eaten raw (e.g. most fruit and vegetables once they have been washed and/or peeled).
- Other foods are cooked to improve their texture and flavour (e.g. potatoes) or to kill harmful (pathogenic) bacteria that could cause food poisoning.
- Some meat and all poultry must be cooked right through to the middle to ensure the bacteria are destroyed and the food is safe to eat.
- Minced meat products such as sausages and burgers should always be cooked right through as the mincing process mixes the bacteria on the outside of the meat deep inside the food.

Temperature control of food

Temperature probes may be used for:
- checking to see if food is properly defrosted before cooking (e.g. poultry, joints of meat)
- checking to see if food is cooked to a safe temperature (75°C or above)

- checking to see whether ready-to-serve food is still piping hot
- checking to see whether reheated food has reached a temperature of at least 75°C (reheat food once only).

How to safely use a temperature probe

- To check your temperature probe is working properly, crush some ice in a beaker with a little cold water, the temperature should read between -1°C and 1°C. Then test it with the steam of a boiling kettle, it should be between 99°C and 101°C.
- Clean and disinfect the probe before and after each use (anti-bacterial wipes are useful for this, but only use once).
- Switch on; the probe should display the room temperature (around 17–20°C).
- Insert probe into the thickest part of the food; don't let it touch a bone or the base of the tin.
- Wait for the gauge to settle before you take the reading.
- The temperature should reach 75°C or higher.
- If the required temperature is not reached, cook food for longer and then repeat the test.

Defrosting food

- Some foods are best cooked from frozen (e.g. vegetables, fish, burgers).
- Bulky foods such as joints of meat and poultry should be defrosted first to ensure they cook in a reasonable time and evenly, and to achieve the core temperature of 75°C needed to destroy pathogenic bacteria.
- With poultry, there should be no ice crystals, and the legs and wings should move fairly freely.

Reheating food

- The internal temperature when reheating should reach 75°C to prevent food poisoning.
- When using the microwave to reheat, turn or stir the food to ensure even heating and to avoid 'cold spots' in the food.
- Keep food in the fridge until you are ready to reheat it.
- Keep food covered and handle as little as possible.
- Divide food into smaller quantities so the reheating time is reduced.
- Serve reheated food straight away.
- Don't reheat cooked food more than once.
- Throw away any reheated food that is not eaten.

Care with high-risk food

REVISED

- Foods that provide bacteria with moisture, a high protein content and a neutral pH provide ideal conditions for bacterial growth, especially in a warm environment. Foods that provide these conditions and are ready to eat without further heating or cooking are known as high-risk foods.
- High-risk foods are the most likely to cause food poisoning. Extra care should be taken when preparing and cooking foods that are high risk. See pages 61 and 63 for more on micro-organisms and enzymes.

Figure 7.6 This wrap with meat and mayonnaise is a high-risk food

Serving food

- If possible, serve food as soon as it has been cooked; this will ensure bacteria have no time to multiply.
- If food needs to be kept hot, it should be kept piping hot until it is ready to be eaten.
- Food that needs to be chilled or frozen should be cooled as quickly as possible before this is done.
- Never put hot food in the fridge; cool down to room temperature first within 90 minutes.
- Food should be served with clean utensils such as tongs to avoid cross-contamination. Avoid touching food with your hands if possible

Typical mistake

You may be asked to describe ways to avoid cross-contamination. A common exam mistake is to write 'keep raw and cooked foods separate' or 'don't let raw and cooked foods touch one another'; you will gain limited marks for this as it could apply to a meal of cooked chicken with salad. Instead think about naming specific foods. For example:
- keep raw chicken away from cooked meats (e.g. use different-coloured chopping boards for preparing raw and cooked meat)
- use different utensils for handling raw and cooked meat (e.g. at a barbecue have two sets of utensils for putting food on and taking it off the barbecue).

Exam tip

You should learn the rules for personal hygiene and be able to explain the reasons for them.

Now test yourself

1 What is the most common way in which bacteria are spread onto foods? [1 mark]
2 Name three occasions when you should wash your hands. [3 marks]
3 Explain why it is not safe to eat a rare (undercooked) beef burger. [2 marks]
4 When using a temperature probe, why should it be inserted into the thickest part of the food? [2 marks]
5 Why should any food that has been reheated be thrown away if it is not eaten? [3 marks]

8 Factors affecting food choice

Factors that influence food choice

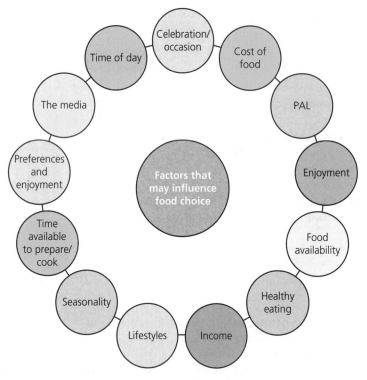

Figure 8.1 **Factors that may influence food choice**

Physical activity level (PAL)

REVISED

- How active you are may influence your choice of food.
- Your **PAL** shows your daily activity level as a number.
- It is important that food intake is balanced with energy expenditure to maintain a healthy weight.
- If you are not very active (sedentary) you will have a lower PAL than someone more active.
- The PAL and your **basal metabolic rate (BMR)** can be used to work out how much food energy you need to consume, and this may affect your food choice.

Celebration/occasion

REVISED

- Food can play a vital role in any celebration or special occasion, such as a birthday.
- Religious festivals during the year, such as Easter, Christmas and Rosh Hashanah, are often associated with specific foods.

Figure 8.2 A celebration meal

Cost of food

REVISED

- Shops now offer food at a wide variety of price ranges, from value ranges (low cost) to premium ranges (high cost).
- Foods are often cheaper in a supermarket than in a corner shop.
- Discount food retailers such as Aldi and Lidl can offer very competitive pricing on selected products.

Enjoyment and preferences

REVISED

- We choose food we like to eat because it provides enjoyment and meets an emotional need.
- The smell, taste, texture and appearance of food stimulate all of the senses.
- Everyone has unique likes and dislikes. These preferences develop over time, and are often influenced by personal experiences.

Food availability

REVISED

Much of our food is available all year round. The choice has increased in recent years due to new developments in:
- transport
- preservation
- storage of foods.

Healthy eating

REVISED

- A wide range of reduced-fat, low-calorie, sugar-free and salt-free food products now exist due to greater public awareness of the potential risks to health of a high-fat, high-salt and high-sugar diet.
- The increase in the number of consumers participating in slimming diets, and following specialist, low-carbohydrate or high-protein diets, has also widened the choice of products available.
- Interest in healthy convenience foods has grown.

Income

REVISED

- Those with a higher disposable income spend more money on ready-to-eat 'premium' food products and higher-quality food with minimal preparation.
- More high-fat and high-sugar foods may be chosen if income is limited.
- Cheaper sources of protein (e.g. pulses, tinned fish, eggs) and cheaper types of fish (e.g. coley) may be purchased by families on a lower income.

Lifestyles

REVISED

- People have busy and flexible lifestyles. More women and men are now working full-time, reducing the time and motivation they have to cook meals every evening.
- Working parents may choose to eat out, or purchase ready meals or part-prepared food products.
- As the number of people living alone has increased, the number of single-portion ready meals purchased and consumed has grown.
- Many people now travel greater distances to work. This means they will often eat on the move.
- Food products reflect our more flexible lifestyles (e.g. protein shakes that can be consumed quickly).
- Due to busy lifestyles within families, there is often less emphasis on eating family meals together. Family members may have activities in the evenings, so quick snacks or ready meals are more convenient, and family members may not all eat together every day.

Seasonality

REVISED

Some foods are seasonal, which means they are available only at certain points in the year. People choose to eat seasonal foods because they can be:
- plentiful, and therefore often cheaper
- locally produced, so fewer food miles
- fresher.

In the UK, developments in the transport, preservation and storage of foods mean that much of our food is available all year round, although it may be more expensive to buy when it is out of season.

Time of day

REVISED

- We tend to choose different foods to eat in the morning for breakfast, in the middle of the day at lunchtime and in the evening for dinner/tea/supper.
- Snacks may be eaten between meals.

Time available to prepare and cook

REVISED

- Due to people's lifestyles being so busy there is less time available to prepare and cook meals.
- Consumers are demanding greater convenience from food products, so may buy things like grated cheese, prepared salads and pre-chopped vegetables, or ready-made 'meal deals' for lunch or dinner.

Revision activity

Using an A3 sheet of paper, write the heading 'Factors that may influence food choice'. Copy Figure 8.1 in the centre of the page.

On the sheet, around the diagram, write an explanation of what each factor means.

Costing recipes

REVISED

It is important to know how much a recipe is going to cost to make.

For example, a recipe requires 75 g of Cheddar cheese. Cheddar cheese costs £4.25 for 350 g. To work out the cost of 75 g of Cheddar cheese:
- £4.25 is divided by 350 g to get the amount for 1 g
- this is then multiplied by the amount used, 75 g
- the cost of 75 g of Cheddar cheese is £0.91.

You may also need to work out the cost per portion. So, for example, if a dish serves four people, and you need to work out the cost per portion, you need to divide the total cost of the dish by four.

Typical mistake

Calculate costings carefully, as an incorrect costing of one ingredient can have an impact on the overall cost. Check portion size carefully too, before dividing the cost of the dish as that also has an impact.

Making modifications to a recipe to decrease the cost

- Use cheaper protein foods (e.g. brisket of beef instead of rump steak).
- Change the method of cooking (e.g. stir-fry instead of bake).
- Buy loose produce, not pre-packed (e.g. apples).
- Use value products rather than branded or premium-range products (e.g. value baked beans).

Now test yourself

TESTED

1 Identify two reasons why the choice of food in the UK has increased. [2 marks]
2 State one advantage of eating foods in season. [1 mark]
3 Define the term PAL. [1 mark]
4 Explain three factors that may influence what we choose to eat. [6 marks]

Food choices

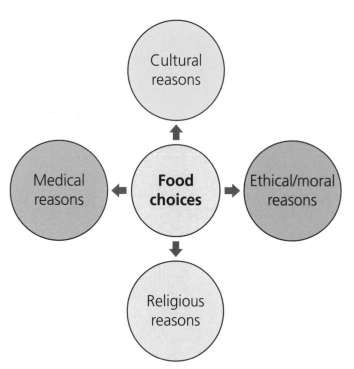

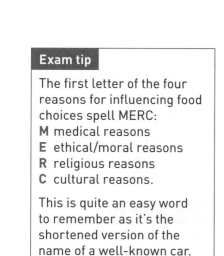

Figure 8.3 **Factors that may influence food choices**

Exam tip

The first letter of the four reasons for influencing food choices spell MERC:
M medical reasons
E ethical/moral reasons
R religious reasons
C cultural reasons.

This is quite an easy word to remember as it's the shortened version of the name of a well-known car.

Cultural reasons for making food choices   REVISED

Culture describes our laws, morals, customs and habits, and these have an influence on what and why we choose to eat. Many cultural groups have guidelines regarding acceptable foods, food combinations, eating patterns and eating behaviours.

Ethical and moral reasons for making food choices   REVISED

People want to have information on the production methods, the ingredients, the country of origin, and how far the foods they buy and eat have travelled (food miles).

Some ethical reasons for food choice include not buying food because:
● an animal has been killed (vegetarians)
● the food has been intensively farmed in poor welfare conditions
● the food has been genetically modified
● too many chemicals have been used in its production
● the food has high food miles and therefore a large carbon footprint.

Religious reasons for making food choices   REVISED

Many religions have specific rules relating to food, and have celebrations and festivals where specific foods are eaten at specific times.

Table 8.1 Religious reasons for food choices

Religion	Foods not eaten	Festivals and celebrations
Jews (Judaism)	Jews do not eat shellfish or pork They do not eat dairy and meat in the same meal They eat meat that has been slaughtered in a specific way in order to be called kosher Kosher means that a food is allowed to be eaten because it is considered clean	Passover is celebrated by a meal and the eating of matza Rosh Hashanah is the Jewish New Year, where a meal is eaten and apples are dipped in honey Yom Kippur is a day of fasting and prayers, when families eat before the sun sets and then fast for 24 hours Hanukkah is the festival of lights, where a lot of food is eaten to celebrate, including fried foods
Hindus (Hinduism)	Hindus do not eat beef or any beef products – they consider the cow to be a sacred animal They will use milk because no animal is killed during its production process Many Hindus are vegetarians, which comes from the principle of Ahimsa (not harming) Most Hindus don't drink alcohol	Hindus celebrate Diwali – the festival of lights – by exchanging sweets
Sikhs (Sikhism)	Sikhs do not eat beef or any beef products because they consider the cow a sacred animal Many Sikhs are vegetarians Many Sikhs will not eat halal or kosher meat, as they believe the animals are not killed humanely Devout Sikhs do not drink alcohol	Sikhs celebrate Guru Nanak's birthday, and this is when they eat their sacred pudding, Karah Parshad, made from equal parts butter, sugar and flour
Muslims (Islam)	Muslims do not eat pork They will eat only halal meat, which has been slaughtered in a very specific way They won't eat seafood without fins or scales Many Muslims do not drink alcohol	Muslims have a month-long fast called Ramadan, during which they don't eat during daylight hours; to celebrate the end of Ramadan, they have a three-day festival called Eid where special food is eaten
Buddhists (Buddhism)	Most Buddhists try to avoid intentionally killing animals and are vegetarians Monks and nuns are usually very strict, and some monks fast in the afternoon	Buddhists celebrate Wesak, where they eat only vegetarian food and don't drink any alcohol
Rastafarians (Rastafarianism)	Eat food that is natural, pure, clean or from the earth (e.g. vegetables); this is called I-tal They try to avoid food that has been chemically modified or contains artificial additives They do not eat pork They eat fish, but will not eat fish more than 12 inches long Food is prepared without salt	Rastafarians celebrate Christmas, when they eat a large feast; the food eaten is mostly vegetarian or vegan, in keeping with their food laws
Christians (Christianity)	Before Easter many Christians will observe Lent, where they give up certain foods for a period of 40 days and 40 nights	Some Catholics fast on Fridays and during the run-up to Lent; Christmas and Easter are times of celebration when traditional foods are eaten

Food choices linked to intolerances and allergies

Food intolerances

Some people have a sensitivity to certain foods, which can give them symptoms such as nausea, abdominal pain, joint aches and pains, tiredness and weakness. This is called a **food intolerance**.

Two examples of food intolerances are lactose intolerance and gluten intolerance.

Figure 8.4 lactose-free products

Table 8.2 Food intolerances

Lactose intolerance	People with lactose intolerance cannot digest the sugar in milk, called lactose
	As they cannot digest lactose, it gets broken down in the stomach by bacteria, which causes abdominal pain, nausea, diarrhoea and flatulence
	They avoid all dairy products.
	There are alternatives to dairy, such as: soya milk and soya products; and lactose-free products.
Gluten intolerance/ coeliac disease	Coeliac disease is a bowel disease; it is an intolerance of gluten
	Gluten is a protein present in a number of cereals, including wheat, rye, oats and barley
	A person suffering from coeliac disease is called a coeliac
	Symptoms include: ● weight loss ● diarrhoea ● lack of energy ● loss of appetite and vomiting ● children may not gain weight or grow properly ● general malnutrition, as a coeliac cannot absorb enough nutrients
	A coeliac needs to follow a strict gluten-free diet, avoiding wheat, flour, baked products, bread, cakes, pasta and breakfast cereals
	Many foods are naturally gluten-free and can be eaten freely (e.g. fresh vegetables, meat, fish, cheese, eggs)
	Gluten-free flour and gluten-free products are readily available these days

Food allergies

A food **allergy** is an allergic reaction to a specific food. Foods that may cause an allergic reaction include shellfish, nuts, eggs, milk, wheat and fish.

● Someone with an allergy will need to take great care that they do not eat these foods
by mistake
● If they do eat food they are allergic to they may experience an itchy sensation inside the mouth, throat or ears, or swelling of the face, around the eyes, lips, tongue and roof of the mouth; in some cases it

can bring about anaphylactic shock, which could be fatal unless an Epipen containing adrenalin is used
- It is crucial that all food labels and food eaten outside the home are checked very carefully
- Be aware of what others around are eating – allergic reactions can be triggered by just being near someone eating peanuts; for instance

Now test yourself

TESTED ☐

The ingredients listed below are for a cheese sauce.

Cheese sauce
- 50 g margarine
- 50 g plain flour
- 500 ml milk
- 100 g grated cheese

1 Why is the sauce not suitable for a coeliac?   [1 mark]
2 Why is the sauce not suitable for a person who is lactose intolerant?   [1 mark]
3 Name the religion that uses milk because no animal is killed during its production process?   [1 mark]
4 Why could a Jew not eat the cheese sauce with some grilled steak?   [1 mark]

Revision activity

Make a set of ten revision cards for each of the religions, and each of the medical conditions. Include some information on foods the people in question can eat and foods they avoid.

Food labelling and marketing influences

Food labelling

REVISED

Food labelling contains information provided by food businesses about their products, and covers all food that is sold in shops, cafés, restaurants and other catering establishments.

Food labels provide information for the following reasons:
- to enable the consumer to make informed decisions and choices, and educate them about the food they choose to buy
- to help us store, prepare and cook the food we choose to buy correctly
- to establish the nutrient content of the food
- to identify the fat or sugar content of the food
- if the consumer has a health condition, such as diabetes or high blood pressure, they may want to check the carbohydrate or salt content of the food
- if the consumer has a severe allergy to certain ingredients (e.g. nuts), they need to check if the food contains those ingredients
- if they need to complain about the food, they will need the manufacturer's name and address
- the consumer may want to buy local produce, or be environmentally aware and will want to know where the food comes from.

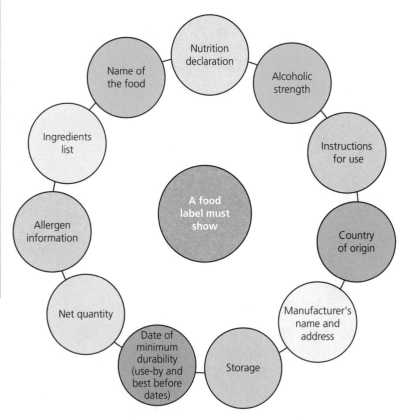

Figure 8.5 The mandatory and consumer information required on a label

From December 2016 nutritional labels on food must contain the following information:

- the energy content of food, stated in kilojoules (kJ) and kilocalories (kcal) per 100 g and per serving
- reference intakes (RIs), which are expressed as a percentage that a serving provides
- saturates, 'sugars' and 'salt'
- energy values, shown in both kilojoules (kJ) and kilocalories (kcal).

The traffic light label

- The traffic light label includes information on the percentage of the reference intake (RI), alongside the traffic light colours.
- The colours are red (high), amber (medium) or green (low) for amounts of fat, saturated fat, salt and sugar.
- Information is given on the types of fats, protein, fibre, carbohydrates and salt in the food or drink.
- This information is given per 100 g and per serving.

List of nutrients

The nutritional information can be found on the back of the label. The nutrients must be listed in the following specific order:

- energy
- fat
- saturates
- carbohydrate
- sugars
- fibre (not required by law)
- protein
- salt
- vitamins and minerals.

Manufacturers can state the nutrition information per portion in addition to the required per 100 g or per 100 ml information.

Nutrition claims

There are stringent rules about making claims regarding nutrition and health, so as not to mislead the consumer. An example of a permitted nutrition claim is **fat free** – this means that the food product contains no more than 0.5 g of fat per 100 g or 100 ml. Manufacturers are not allowed to state a product is 'x% fat-free'.

Any **health claim** a manufacturer makes has to be reviewed to ensure it is accurate before it appears on the label.

Interpreting nutritional labelling

REVISED

- Food labels can help us to understand what nutrients food products contain.
- A red label means the food is high in something consumers should try to cut down on in their diet (sugar, salt and fat) or should have only occasionally; such foods should be chosen less frequently and eaten in small amounts.
- An amber label means the food isn't high or low in the nutrient, so this is an acceptable choice for the majority of the time.
- A green label means the food is low in that nutrient. The more green, the healthier the choice.
- The consumer should choose foods with more greens and ambers and fewer reds, to ensure healthier choices.

Exam tip

Ensure you know why food labelling is important.

1/3 of a pie (oven cooked)

ENERGY 2218kJ 533kcal — 27%
FAT 34.5g — 49%
SATURATES 16.1g — 81%
SUGARS 2.3g — 3%
SALT 1.25g — 21%

% of the Reference Intakes

Typical values per 100g: **Energy 1210kJ/291kcal**

Figure 8.6 The traffic light label

- The label also tells you the number of grams of fat, saturated fat, sugars and salt in what the manufacturer or retailer suggests is a 'serving' of the food.
- Reference intakes (RIs) are on labels so that consumers can understand what their daily intake of specific nutrients should be.

How marketing can influence food choice

Marketing is identifying consumers' needs and wants, and using that information to supply them with products that match their needs and wants. Several methods are used to promote a food product to a consumer.

- Food products are advertised on television, the internet, radio, newspapers and magazines. Brand loyalty can be built upon and a carefully controlled image of the product can be created.
- Buy one get one free (BOGOF) offers
- Price reductions and special offers
- Meal deals (e.g. a main course, side dish, dessert for two people, plus a bottle of wine or non-alcoholic alternative for £10)
- Free samples to taste or try in store
- Product placement
- Loyalty cards – each time the card is used points are awarded to the consumer, which may be converted into 'money off' coupons for use in store or used to purchase goods and services from participating organisations.

Figure 8.7 A loyalty card

The retailer is also able to monitor consumer purchases.

Now test yourself

1 Explain the difference between 'best before' and 'use by' dates. [4 marks]
2 State two ways in which a supermarket can promote a new product range. [2 marks]
3 State three reasons why food labels are important to a consumer. [3 marks]
4 Explain what the red means on the traffic light label that appears on food products. [2 marks]

Typical mistake

If you have an examination question on marketing, read it carefully – is it asking about the consumer or the retailer?

9 British and international cuisine

Cuisine is a style of food characteristic of a particular country or region.

British cuisine

Table 9.1 British cuisine

Distinctive features and characteristics of cooking	**Traditional preparation techniques**: filleting, slicing, mashing, shredding, scissoring, snipping, scooping, crushing, grating, peeling, segmenting, de-skinning, de-seeding, blanching, shaping, piping, blending and juicing
	Traditional foods grown: vegetables, such as potatoes, onions, leeks, Brussels sprouts, peas and beans; fruit, such as apples, strawberries and plums; herbs, such as mint, chives and sage
	Traditional foods reared: beef, lamb, poultry and game, pork, bacon and ham
	Traditional foods caught: fish, such as mackerel, cod, haddock and salmon
	There are also many different variations of sweet puddings and cakes in Britain (e.g. Victoria sandwich cake, shortbread, trifle); each of them usually begins with the same basic ingredients of milk, sugar, eggs, flour and butter; many desserts use home-grown fresh fruit, such as raspberries or strawberries
Equipment and cooking methods used	**Cooking methods**: baking, roasting, casseroling, grilling, barbecuing, braising, steaming, boiling, poaching, simmering and frying
	Equipment: casserole dishes, moulds and tins for roasting and baking
Eating patterns	Many people in the UK will have three meals a day: breakfast, lunch and an evening meal
	Breakfast
	The classic 'full English breakfast' is not normally eaten every day – it is usually a weekend or holiday treat
	Most people in the UK will have cereals, yoghurt and fruit, toast or porridge for breakfast
	Elevenses
	An old-fashioned name for a mid-morning snack
	Usually consists of a cup of tea or coffee and some biscuits
	'Brunch'
	A combined breakfast/lunch meal
	Lunch
	On most days this is likely to be a light, quick meal (e.g. sandwich, soup or salad)
	Usually eaten between 12 pm and 2 pm
	On a Sunday, a traditional Sunday lunch may be eaten (also called 'Sunday roast', because the main dish is roasted meat such as roast beef, pork or lamb)
	Tea
	An afternoon snack (the mid-afternoon version of 'elevenses')
	Afternoon tea is popular in tea rooms and cafés; traditional foods eaten at afternoon tea are small sandwiches, small cakes and savouries

	Dinner Most people eat between 6 pm and 9 pm Often a time to socialise with family or friends after work or school **Supper** Sometimes the evening meal is called supper, but the traditional supper tends to be served a bit later – usually between 7.30 pm and 9 pm Usually home cooked An old-fashioned term – most people now call it dinner
Presentation styles	The classical plating style is often used. This method of styling uses the idea of a plate being a clock: between 12 and 3 are vegetables; between 9 and 12 are starchy foods; between 3 and 6 is the main component of the meal Sauce can be used to present food by pouring it, drizzling it, dotting it or serving it in a small jug alongside the food Garnishes and decorations can be used to enhance a dish Savoury food is often served in oval dishes or on oval plates, and sweet food on round plates or in round dishes Cakes, biscuits or scones are sometimes placed on a doily and then on a plate, or on a tiered cake stand

Figure 9.1 A full English breakfast

International cuisine

Many traditional British dishes have been replaced by dishes that originated in other countries and cultures.

Exam tip

You need to study only British cuisine and two other international cuisines.

Italy

REVISED

Italian flavours are very simple. The diet is based on fresh vegetables, herbs, fish and olive oil.

Table 9.2 Italian cuisine

Traditional foods grown/reared/caught	Northern Italy is cool and mountainous – rice is grown and the land is mostly used to rear animals for cured meats; southern Italy is hotter, so more crops are grown there **Grown**: olives, tomatoes, rice and lemons **Reared**: cured meats (salami, pancetta, Parma ham and prosciutto), cheeses (ricotta, mozzarella, Parmesan and pecorino) **Caught**: fish (tuna, swordfish, shellfish and squid)
Traditional dishes	Gnocchi, pizza, bread (e.g. ciabatta), risotto, pasta and pasta dishes (e.g. lasagne, cannelloni, spaghetti Bolognese, spaghetti carbonara, ravioli), desserts (e.g. tiramisu, pannacotta, arancini, biscotti, ice cream), minestrone soup, drinks (e.g. espresso coffee and cappuccino)
Main cooking methods	Boiling (especially for pasta and gnocchi) and baking
Specialist equipment used for cooking	Pizza and bread are often cooked in a wood-fired oven, or in a traditional oven on metal trays Pasta-making machine Large saucepan for cooking pasta Slotted spoon for removing pasta from pan Gnocchi board – useful for shaping gnocchi correctly
Eating patterns	Italians may have three or more courses to their meals, which are eaten over a long period of time Breakfast is usually light – coffee (either espresso or cappuccino) and some bread rolls Two main meals – one is served in the middle of the day, the other in the evening, usually quite late Traditional meals consist of: starters (antipasto, soup/pasta or risotto); fish, meat or poultry with accompaniments of either vegetables or salad; dessert (dolce or fruit and cheese); coffee at the end of a meal
Presentation styles	Often food is presented very simply; the presentation style could be described as rustic Large (wooden) bowls of food with serving spoons are often used There is an emphasis on sharing

Exam tip

You need to study only British cuisine and two other international cuisines.

Figure 9.2 Traditional Italian dishes

Spain

Spanish cooking is simple and rustic with a focus on seasonal and local produce, including fish and vegetables. Very little dairy food is eaten.

Table 9.3 Spanish cuisine

Figure 9.3 Paella: a traditional Spanish dish

Traditional foods grown/ reared/caught	Northern Spain has heavier rainfall than the south, so it is good for growing crops and rearing animals; southern Spain is hotter – many types of fruit are grown, including grapes for making wine and sherry
	Grown: vegetables (e.g. tomatoes, fresh peppers and chillies called pimentos); herbs and spices (e.g. saffron, cumin, coriander seeds, paprika and smoked paprika); fruits (e.g. grapes, figs, melons, dates and apricots); bread (e.g. pan de cebada); olives (also made into olive oil)
	Reared: meat (e.g. pork, poultry preserved meat – black, red and white sausages, chorizo, hams (jamón and Serrano)
	Caught: fish (e.g. anchovies, shellfish); preserved fish (squid and calamari)
Traditional dishes	Paella, frittata, gazpacho, bacalao, cocas (pastries), tapas and pinchos, stews, churros, magdalenas, polvorones
Main cooking methods	Simmering, stewing and braising
	Roast, fry and sauté
	Grill and barbecue meats and vegetables
Specialist equipment used for cooking	Paellera – the traditional large round pan with shallow sides used for cooking paella
	Garlic crusher – Spanish stews and soups tend to use a lot of garlic
	Ceramic and glazed pots – used for both cooking and serving food; a cazuela is a glazed terracotta dish that comes in different shapes and sizes
	Espresso coffee pot – coffee is served after every meal
	Pestle and mortar – for grinding herbs and spices to make dips for tapas, and to grind saffron and salt together for paella
Eating patterns	Because of the hot climate, the Spanish have breakfast early – usually coffee and rolls
	Lunch is around 2 pm to allow time in the afternoon for a siesta; the meal is usually light, consisting of stew or soup, fruit and then coffee
	The evening meal is eaten much later and is often meat, potatoes, salad, fruit and coffee
	Before the evening meal, the Spanish may eat tapas
Presentation styles	Very much like the Italians, the Spanish often present their food very simply, in terracotta bowls to share – tapas is characeristic of this style.

China

Chinese food is traditionally quick to make, with an emphasis on colour, flavour and texture.

Table 9.4 Chinese cuisine

Traditional foods grown/reared/ caught	China has a rugged, quite barren landscape; its people grow potatoes, tomatoes and aubergines
	In the north of the country, wheat is grown to make noodles
	In the south of the country, rice and soya beans are grown
	There are very few cattle because there is little grazing land
	Grown: noodles and rice, grains, vegetables (e.g. Chinese leaves, water chestnuts, bamboo shoots and beansprouts), fruit (e.g. lychees and kiwi fruit)
	Reared: pork, duck and chicken
	Caught: fish and seafood
Traditional dishes	Spring rolls, soups/broths, dumplings, prawn toasts, beef/prawn chop suey, szechuan pork or beef, crispy duck/Peking duck, fried rice, sweet and sour noodles, chinese food is often flavoured with ginger, garlic and soy sauce
Main cooking methods	Stir-frying, steaming and frying
	In a traditional Chinese kitchen there is no oven
Specialist equipment used for cooking	Wok, chinese cleaver, bamboo steamer, steel spider meshed scoop, clay pots with lids
Eating patterns	Soups are an important part of the Chinese diet; they are not often eaten at the start of a meal but throughout or at the end; soups are also eaten for breakfast
	At mealtimes, the Chinese family will have four or five dishes at the same time; each person has a small bowl with rice in it
	Sweet foods are known as desserts; they are often eaten separately, as part of a meal or as a snack
	The traditional drink with a meal is green tea
Presentation styles	The Chinese use a range of colourful foods
	They prepare vegetables to make them look as attractive as possible
	They tend to have several dishes of food served in small bowls, eating a little of everything with chopsticks and large rounded spoons

Figure 9.4 Chinese serving dish and cutlery

India

India is a vast country made up of several states; there are huge variations in climate and religion – and therefore foods.

Table 9.5 Indian cuisine

Figure 9.5 Traditional Indian dishes

Traditional foods grown/ reared/caught	India is a mainly agricultural country, primarily growing crops
	In the north of India, wheat is the staple food used to make chapattis
	In the south, rice is the staple food and more curries with plenty of sauce are served; there is more rainfall in the south, so many vegetables are grown there
	India has a coastline, so fish is also a major industry
	Grown: wheat, maize, basmati rice, lentils, spices (e.g. cumin, turmeric, chilli, cardamom), tea (e.g. Darjeeling, Assam), vegetables (e.g. aubergines, okra and peppers)
	Reared: goat, lamb and chicken
	Caught: fish
Traditional dishes	Tandoori, naan bread, tikkas, kormas, biryanis, parathas, poppadoms, samosas, onion bhajis, lentil dhal
Main cooking methods	An oven is rarely used in India; food is cooked on hobs
	Indian food is often fried, and a variety of spices are blended and used to enhance the flavour of the food cooked
Specialist equipment used for cooking	Tandoor – a clay oven heated by charcoal used to cook naan bread and tandoori dishes
	Food is often fried in ghee or oil
	Karhai – the Indian version of a wok used to simmer, stir-fry, deep-fry and steam
	Flat griddle pan – used for cooking chapattis and parathas
	Mortar and pestle or mini food processor as many spices are blended together
Eating patterns	Entertaining in the home is very popular
	All the dishes are placed on the table at once and shared
	Indians enjoy snacks – there are many street markets selling food Snack food is also prepared in the home for any visitors
	Desserts are not usually served every day; they are served only on special occasions, such as for the festival of Diwali
Presentation styles	In central India, small dishes of food are served on a tray or a thali – a stainless-steel plate, on which several small dishes are placed around the edge; in the centre, there are pickles, breads, rice and poppadoms

Revision activity

Copy out and complete this table as a revision card:

Country	Specialist equipment used for cooking	Foods grown/ reared/ caught	Traditional dishes	Eating patterns	Presentation styles
Britain					
Italy					
Spain					
China					
India					

Typical mistake

Check the examination question carefully – international cuisine means food and food products from a country outside of the UK.

TESTED ☐

Now test yourself

1 What is the definition of cuisine? [1 mark]
2 Explain two reasons why Chinese food is considered healthy. [4 marks]
3 Describe two pieces of equipment traditionally used in British cuisine. [6 marks]
4 Define elevenses. Identify what is usually eaten or drunk for elevenses. [2 marks]

10 Sensory evaluation

The sensory evaluation of foods, food products and combinations of foods is very important to make sure they are acceptable to consumers.

What is sensory evaluation?

Sensory evaluation is carried out on food products by setting up a taste panel to ensure:
- a food product meets a consumer's expectations
- changes to an existing food product mean the product remains acceptable
- food products remain consistent over time
- a food product compares to other similar products
- a food product meets the original specification
- the quality and shelf life of food products over time.

The characteristics of food affect your senses of sight, taste, touch and smell. These are known as **organoleptic qualities**. Judging food based on these characteristics is called sensory evaluation, or sensory analysis.

When you eat food, you are judging its:
- appearance
- consistency
- taste
- smell, or aroma.
- texture/'mouthfeel'

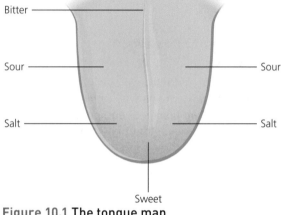

Figure 10.1 **The tongue map**

How the taste receptors work

REVISED

- Taste influences the selection of food to eat. Food must be dissolved in water, oil or saliva to have a taste.
- Taste is detected by taste buds. There are about 10,000 taste buds on the surface of the tongue. The taste buds will tell the brain if the food is sweet, sour, bitter or salty.
- There are four basic tastes that you can recognise:
  1 sweet   3 sour
  2 salt   4 bitter.
- A 'fifth' taste, called **umami**, has been found to have its own taste receptors, and is found in foods such as Cheddar cheese, tomatoes and Japanese miso.

How the smell receptors work

REVISED

- The nose is a vital organ for taste as well as smell. The taste buds on the tongue can detect only sweet, sour, bitter and salt.
- All the other 'tastes' are detected by the **olfactory system**, which is the smell device in the nose. It allows you to identify aromas or smells.
- The smell, or aroma, of food makes a gas. When this gas enters the nose it passes over the epithelium. The epithelium contains more than 10 million receptors and is able to identify different smells.
- When we discuss food we often refer to its flavour. Flavour is the combined sense of taste, mouthfeel and aroma.

> **Exam tip**
>
> Make sure you understand the specific purpose of the each of the sensory analysis tests.

Different types of sensory analysis tests and their uses

Preference tests

Preference tests are sometimes called acceptance tests. They are used to find out if a food product is acceptable to the consumer.

The main types of preference test are:

- **paired preference test** – the tester is given two samples of a food product and asked, 'Which sample do you prefer?'
- **hedonic ranking** – a preference test used to find out how much people like or dislike a product.

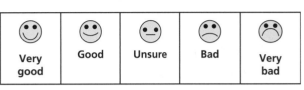

| Very good | Good | Unsure | Bad | Very bad |

Figure 10.2 Example of a five-point hedonic scale

Discrimination tests

- Discrimination tests are used to detect differences between two samples.
- The **triangle test** is a discrimination test to find out whether small differences between two products can be detected. Three samples are presented to the tester. Two of the samples are identical and one is different. The tester is asked to identify the 'odd one out'.

Grading tests

These tests measure the strength of a specific sensory property.

The main types of grading test are:

- **ranking test** – used to measure the strength of a specific sensory property in a number of samples (e.g. sweetness in biscuits); the tester will rank the samples in order – the sample that is strongest in the specified sensory property will be ranked first
- **rating test** – allows people to rate the extent to which they either like or dislike one aspect in a number of similar food products; this could be, say, the spiciness of a range of samosas
- **profiling** – used to obtain a detailed description of the appearance, taste and texture of a food product; features such as mouthfeel, appearance, aroma, flavour and texture can be assessed separately or together in one profile; the results of the profiling test are plotted on a star diagram to give a visual profile of the product – this is called a sensory profile.

> **Exam tip**
>
> Thorough revision of the different tests is crucial so as not to get them muddled in a response to an exam question.

Revision activity

Copy and complete the table below, leaving yourself enough space to write your answers in the second and third columns. The first row has been completed for you.

Sensory analysis test	Use	Examples
Preference test	To find out if a food product is acceptable to the consumer	'I have made a low-salt lasagne and want to see if it is acceptable to consumers – I will ask at least five people to taste the lasagne to obtain reliable results'
Paired preference test		
Hedonic ranking		
Discrimination test		
Triangle test		
Grading tests		
Ranking test		
Rating test		
Profiling		

How to carry out a sensory evaluation

The testing environment

- The testing room should be a controlled environment; there should be no distractions.
- The lighting should be controlled.
- The seating should isolate the testers from one another; individual testing cubicles are ideal.

Figure 10.3 Setting up the testing

The food testers

- Untrained consumer testers can be used to taste-test foods.
- Trained food tasters can be used when it is necessary to identify subtle differences between food products.
- All food testers should be in good health, non-smokers and free from any illness such as colds.
- The testers must not have strong likes or dislikes for the food being tested.

The food samples

- The samples should all be of the same size; this is usually enough for two bites or sips.
- Samples should be labelled with random three-digit numbers; the numbers 1, 2, 3 or letters A, B, C are not used because they could influence decision making.
- The holding time before the sample is tested should be monitored. Some products can deteriorate if held for too long before testing.

The testing equipment

- The plates, spoons or cups used to present the samples should be an identical shape, colour and size.
- Food carriers such as bread, crackers or pasta are sometimes used if the food product is usually eaten with another food.
- Water should be provided to cleanse the mouth in between each sample. The tester should also have a 30-second rest period between tasting each sample.

Now test yourself

TESTED

1 State two reasons why you would carry out sensory analysis. [2 marks]
2 Name five basic tastes your tongue can recognise. [5 marks]
3 Describe when you would use a difference test, such as a triangle test. [2 marks]
4 Describe how you would carry out a sensory analysis to compare a bought spaghetti Bolognese with a home-made one. [6 marks]

11 Environmental impact and sustainability of food

Food sources

Free-range food production

REVISED

- This is a method of farming where animals have access to outdoor spaces for at least part of the day.
- Animals farmed in this way include pigs, grass-fed beef, laying hens, chickens and turkeys.
- Free-range does not mean organic.

Table 11.1 **Advantages and disadvantages of free-range production**

Advantages of free-range production	Disadvantages of free-range production
Free-range animals can graze and look for food The animals are more likely to behave naturally and eat a more varied diet	Free-range animals can be at risk from being attacked by predators – for example, foxes will hunt chickens
Some consumers feel that animals produced by free-range production have had a better life, and therefore prefer to buy these products	Free-range animals can catch diseases from wild animals – some people believe that badgers spread diseases to cows
Some consumers and chefs believe that the meat from free-range, grass-fed cows tastes better than that from intensively farmed grain-fed cows	Free-range animals can suffer discomfort during extreme weather – for example, during heavy snowfall sheep can become trapped and die
Free-range animals can roam around and are less likely to spread diseases among themselves	More land is needed for free-range production and it needs to be carefully managed; this means that free-range foods can be more expensive to buy

Intensive farming

REVISED

- This is a farming system that aims to produce as much yield as possible, usually with the use of chemicals and in a restricted space.
- **Intensive farming** can be used with both crops and animals.
- Intensive production means that animals can suffer from isolation or overcrowding, and cannot move around or behave naturally.
- Animals can be restricted from natural behaviours like grazing, foraging, mud wallowing, running and nesting.

Figure 11.1 **Free-range hens**

Genetically modified (GM) foods

- **GM food** is produced from plants that have had their genetic make-up changed by scientists.
- Scientists can 'cut and paste' genes from one plant to another to give a plant new features (e.g. making it able to resist drought).
- In the long term, GM food could mean cheaper and more plentiful food supplies, but there are concerns that we do not know the consequences of meddling with nature.
- At present, the production of genetically modified food in Europe is restricted.

Advantages of GM foods are:
- Better resistance to insects, pests or disease.
- Increased storage life when harvested.
- Can survive poor weather conditions (e.g. resistance to low rainfall).
- Faster growth.
- Can be produced in large amounts.
- Cheaper as don't need as much pesticide and herbicide.
- Fewer people needed to grow them.
- Can be developed to have a large amount of a specific nutrient in them (e.g. 'Golden Rice' has been produced to contain high levels of vitamin A).

Disadvantages of GM foods are:
- The pollen from GM crops could mix with wild plants, affecting the natural species in the long term.
- The labelling of GM foods can cause confusion: products with more than 1% of GM ingredients need to be labelled; products such as meat, milk and eggs from animals fed on GM animal feed do not need to be labelled.
- Could affect animal habitats.
- The food source for an animal could change if a new plant is introduced.
- Pests and weeds are developing resistance to pesticides and herbicides, so more may be needed.
- The loss of weeds in farmers' fields means the loss of food and shelter for animals and insects.
- The side effects of eating GM foods such as Golden Rice over a long period of time are not known and this is a concern.

Now test yourself

TESTED

1	State three advantages of free-range production.	[3 marks]
2	What is meant by the term GM foods?	[1 mark]
3	Explain three advantages of GM food.	[3 marks]
4	Explain three disadvantages of GM food.	[3 marks]
5	Define intensive farming.	[1 mark]
6	Define free-range production.	[1 mark]

Food and the environment

To sustain our environment, we need to
- use less energy
- reduce the consumption of water
- avoid waste
- recycle and reuse as much as possible.

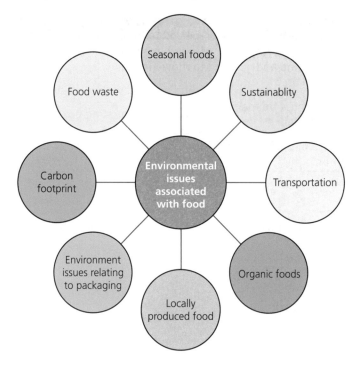

Figure 11.2 **Environmental issues associated with food**

> **Exam tip**
>
> Learn all the environmental issues associated with food by making a copy of Figure 11.2, and then extending it by writing examples and explanations alongside each issue.

Seasonal foods

Seasonal foods are available only at certain points in the year.

Advantages of using seasonal foods are:
- they are more likely to be local or grown in the UK
- the food miles will be low
- they will support local farms
- they are often healthier
- they are plentiful and therefore cheaper.

Figure 11.3 **Seasonal foods**

Sustainability and sustainable methods of farming

Many farmers now work in a sustainable way; this means that they grow crops, or rear animals, in a way that maintains and improves the environment.

RSPCA Assured

RSPCA Assured, previously known as Freedom Food, is the RSPCA's food label. The label indicates that the animal has had a good life, and has been treated with compassion and respect.

The RSPCA welfare standards include:
- more space
- natural lighting
- comfortable bedding
- environmental enrichment (e.g. objects for birds to peck at)
- shade and shelter.

Table 11.3 Three types of egg production

Enriched caged production	Free-range egg production	Barn systems
Intensive method of farming The cages typically hold 90 hens and are stacked on top of one another in tiers Cages must provide 600 cm² of useable space per hen	Hens have continuous daytime access to open-air runs and perches The open-air runs must be mainly covered with vegetation Each hen has 4 m² of outside space at all times	Hens have freedom and space to move around Hens are still within a building Perches and nesting boxes are provided

Fish

Fish and fish products with a Marine Stewardship Council (MSC) label:
- come from a sustainable fishery
- ensure that appropriate fishing methods are used
- ensure the supply is maintained and supported.

Three examples of MSC sustainable fish are cod, plaice and salmon.

Transportation

REVISED

Food miles are the distance that food is transported as it travels from producer to consumer.

Food travels much further than it used to because of the demand for:
- seasonal food all year round
- processed food
- cheap food
- a wider range of ingredients from different international cuisines and cultures, which are not produced or grown in the UK.

Organic food

REVISED

- **Organic** food means that at least 95 per cent of the ingredients come from organically produced plants and animals.
- All food sold as organic must be clearly labelled to show that it has met the necessary standards.
- The food must not have any artificial colourings or sweetener.
- No artificial fertilisers can be used.
- Pesticides are very restricted.
- Animal welfare is vital and the animals are always free range.
- The routine use of many drugs and antibiotics is banned.
- Genetically modified (GM) crops and ingredients are also banned.

Organic foods are thought to taste better because no fertilisers or pesticides are used.

The Soil Association promotes organic farming and the reduced use of pesticides.

The reasons for buying locally produced food

- Shopping locally saves on fuel costs.
- Buying locally supports local farmers and producers.
- Reducing food miles can reduce carbon dioxide emissions.
- Buying meat and poultry from a local butcher means that the meat is fresher, it has not been over-packaged and it has not travelled as far, which is better for animal welfare.
- Freshly picked fruit and vegetables are better nutritionally because they are fresher and contain more water-soluble vitamins, which are lost during storage.

Food waste in the home, during food production and retailing

- There are two main reasons why we throw away good food: we cook or prepare too much; and we don't use food up in time.
- The foods we waste the most are fresh vegetables and salad, drinks, fresh fruit, and bakery items such as bread and cakes.

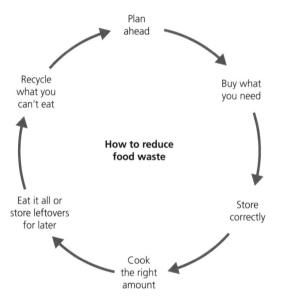

Figure 11.4 How to reduce food waste

- Any food that has been left over can be:
  - recycled as compost, or
  - reused to make another dish (e.g. using leftover chicken to make a curry).

Reducing food waste

- There is now a wider range of different-sized products available, so that the right amount can be bought and not wasted (e.g. 600 g loaves of bread).
- Date labels are now clearer.
- Some supermarkets provide storage advice on their free loose-produce bags.
- Heinz launched a Fridge Pack for its baked beans; they can be kept in the fridge for up to five days after opening, and the snap packs are handy single-person portions.
- Manufacturers offer resealable packs for products.
- Some supermarkets have updated their guidance on freezing.
- Most products now include storage guidance.
- Some packets now carry portion measuring marks on the side to help us cook the amount we need.

How to reduce packaging waste

- Packaging can be recycled or reused.
- Some fruits, such as strawberries, are now in packaging that acts as an 'ethylene remover'. Ethylene is what causes fruit to ripen.
- Some meat is vacuum packed, extending its shelf life.
- Milk is put into white plastic bottles to protect it from light.
- Split packs (e.g. of ham and bacon) give the option of not opening one half so it stays fresher, or freezing one half if we don't want to use it straight away.
- Fruit and vegetables can be bought loose or with little packaging.
- Try to avoid disposable items.
- Either buy only what you need, or buy in bulk and store the excess carefully.
- Look for resealable packs.
- Avoid supermarket plastic carrier bags and take your own.
- Recycle as much packaging as you can, such as glass, paper, cardboard, plastics and cans.

Carbon footprint

REVISED

Our **carbon footprint** is a measure of the impact our activities have on the environment in terms of the amount of greenhouse gases produced through carbon dioxide emissions.

We are advised to reduce our carbon footprint wherever possible by:
- buying local produce and reducing food miles
- cooking fresh meals
- using seasonal ingredients
- cutting down on meat consumption.

Now test yourself

TESTED

1 What is meant by the term carbon footprint? [1 mark]
2 Suggest three ways families can reduce their carbon footprint when buying food. [3 marks]
3 State two RSPCA welfare standards. [2 marks]
4 Explain three ways in which families can reduce food waste. [6 marks]
5 Identify three ways in which we can sustain our environment. [3 marks]
6 State two ways in which you can recycle leftover food. [2 marks]

Sustainability of food

The food security problem

There are four features of **food security** (see Table 11.4). Food security requires all four to be met at the same time.

Table 11.4 The four features of food security

The physical AVAILABILITY of food	Food availability is about the 'supply side' of food security
ACCESS to food	Access to food is affected by the cost of food
The USE of food	How the body uses food and the nutrients in food, and how to eat a balanced diet
The STABILITY of the food supply	Food stability is about the supply of food over time

The types of food insecurity

Table 11.5 Types of food insecurity

Types of food insecurity	Happens when ...	Caused by ...	Can be prevented by ...
Short term	people do not have enough to eat for a short period	a sudden drop in the harvest or lack of access to food due to price increases.	careful planning for possible shortages
Long term	people do not have enough to eat for a long period of time	people living in poverty	long-term action to tackle poverty

- The amount of food produced by a country is called its self-sufficiency.
- The UK is self-sufficient for about 60 per cent of the food we need.
- The remaining 40 per cent of the food we need comes from imports.
- Enough food is produced globally to feed all the people in the world. However, nearly 1 billion people are suffering from hunger and food insecurity.
- In the developed countries, such as those in Europe, the USA, Japan and Australia, food tends to be available and fairly affordable.
- In the less developed countries, such as parts of Africa and Asia, there are people who go hungry because food is expensive and not always available.
- Hunger and food insecurity happen because of poor food distribution and increases in the price of food.

Figure 11.5 Food security is about having enough access to food

> **Typical mistake**
>
> Always read the examination question carefully. If there is a question on this topic, is it asking about food security or food insecurity?

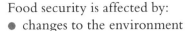

Challenges to providing a secure food supply

Food security is affected by:
● changes to the environment
● population
● trade.

Environmental challenges

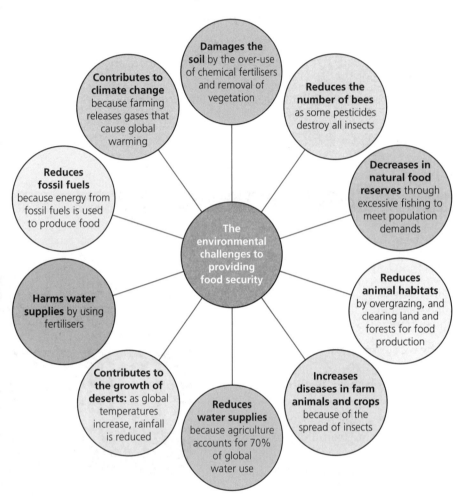

The environmental challenges to providing food security

Damages the soil by the over-use of chemical fertilisers and removal of vegetation

Reduces the number of bees as some pesticides destroy all insects

Contributes to climate change because farming releases gases that cause global warming

Decreases in natural food reserves through excessive fishing to meet population demands

Reduces fossil fuels because energy from fossil fuels is used to produce food

Reduces animal habitats by overgrazing, and clearing land and forests for food production

Harms water supplies by using fertilisers

Increases diseases in farm animals and crops because of the spread of insects

Contributes to the growth of deserts: as global temperatures increase, rainfall is reduced

Reduces water supplies because agriculture accounts for 70% of global water use

Figure 11.6 Environmental challenges to providing food security

Population challenges

● It is expected that the world population will rise from about 7 billion today to 9 billion by 2050.
● Food production in the developing world will need to double to meet demands. As a country develops, it needs more land for housing, which might have been used for growing crops or rearing animals.
● Africa will suffer the greatest insecurity over its food supply if this trend continues, because it also suffers from the effects of climate change and conflicts over land.

Global trade challenges

● Food is traded across the globe.

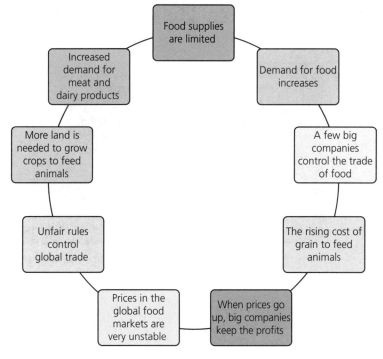

Figure 11.7 Reasons for food price rises

Measures to support local and global markets and communities

To feed the growing population and protect the environment, the production, processing and distribution of food must try to: use the same or less land; use less water, fertiliser and energy; produce less waste.

● Advances in technology can help more people to have access to a wider range of safe, nutritious foods at a reasonable price (e.g. precision farming – involving the use of satellite systems and unmanned aerial vehicles).
● Reducing food waste and packaging saves energy, money and natural resources.
● Eating more vegetables, fruit, cereals and smaller amounts of animal products will help reduce greenhouse gas emissions.
● By choosing fish only from sustainable sources future generations will be able to eat fish and seafood.
● Food security is increased by buying local and seasonal foods. The UK should produce more of its own fruit and vegetables, and consumers should be encouraged to buy them.
● Buying food from trusted sources, offers disadvantaged producers a fair and stable price for their products, and often protects the wildlife and ecosystems.

Now test yourself

TESTED

1 Explain the causes of short-term food insecurity and long-term food insecurity. [4 marks]
2 Explain three environmental challenges in providing food security. [6 marks]
3 Identify three reasons why food prices may rise. [3 marks]
4 State two measures you can take to support the local market and community. [2 marks]

Food production

- Our food supply comes from living plants and animals.
- The two stages of food production are primary and secondary food processing.
- **Primary processing** is changing raw food materials into food that can be eaten immediately or be processed further into other food products.
- It includes the transporting, sorting, cleaning, blending, cooking, preserving, packing and storage of the raw food.
- **Secondary processing** is when primary products are changed into other types of food product.
- Secondary processing includes:
  - wheat flour made into breads, pasta, pastries and cakes
  - milk changing into cheese and yoghurt
  - fruit made into jams and marmalades.

Figure 12.1 A combine harvester collecting wheat so that primary processing can begin

Primary processing: meat and fish

Meat

- Meat and poultry are the muscle tissues of animals and birds.
- Animal meats are beef, lamb, pork, venison, goat and rabbit.
- Poultry meats are chicken, turkey, duck and game birds (e.g. pheasant).
- Offal is the internal organs of animals (e.g. liver).
- Muscle tissue is made up of long, thin fibres held together in bundles by connective tissue.
- There are two types of connective tissue:
  - **collagen** holds the muscle fibres together
  - **elastin** binds the muscle to the bone.
- Long, slow, moist cooking changes collagen to **gelatine**, a soluble protein that is tender and easy to chew.
- Primary processing of beef involves the removal of the bones and jointing the carcass.
- The joints of beef are called cuts.
- Steaks, topside and rib have very little connective tissue and can be cooked quickly by grilling, frying or roasting.
- Shin, brisket, chuck and oxtail have more connective tissue and require slow, moist cooking methods (e.g. braising and stewing).

Fish

- Fish can be grouped according to where they live – freshwater or sea fish – or according to their type – oily, white or shellfish.
- **Oily fish** includes mackerel, salmon, tuna and trout.
- **White fish** includes cod, haddock, plaice and coley.
- **Shellfish** includes crab, oysters, mussels, lobster, shrimps and prawns.
- Fish is tender because it has is no elastin and only a small amount of connective tissue.
- Fish is sold whole or as steaks, tail pieces, fillets and cutlets.
- Fish processing methods include salting, smoking, marinades, pickling, freezing, chilling and canning.

Revision activity

Copy the diagram in Figure 12.2 and add an explanation of how each processing method is used on meat.

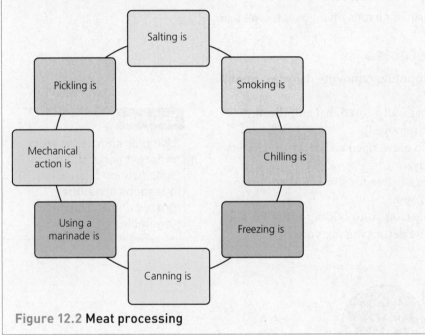

Figure 12.2 **Meat processing**

Primary processing: fruit and vegetables

REVISED ☐

Fruit

- Fruits have three parts: skin, flesh and seeds.
- Citrus fruits include oranges, lemon, grapefruit, limes and satsumas.
- Tree fruits are firm and have edible skin (e.g. apples, pears).
- Stone fruits contain a hard seed (e.g. plums, peaches, cherries, mangoes).
- Berries and soft fruits include strawberries, raspberries and blackberries.
- Dried fruits have had their water removed (e.g. currants, sultanas, dates, raisins).
- Exotic fruits do not grow naturally in the UK (e.g. bananas, passion fruit, kiwi fruit, melons, lychees, pineapple).

Vegetables

- Vegetables are edible plants.
- The different parts of plants produce vegetables (see Table 12.1).

Table 12.1 **Parts of plants that produce vegetables**

Part of plant	Vegetables
Roots	Carrots, swedes, parsnips
Tubers	Potatoes, sweet potatoes, yams
Bulbs	Onions, leeks, garlic
Stems	Celery, asparagus, rhubarb
Leaves	Cabbage, lettuce, spinach
Flower head	Cauliflower, broccoli
Fungi	Mushrooms
Fruits	Tomatoes, pumpkins, cucumbers
Seeds and pods	Pod peas, okra, runner beans, mange tout, sweetcorn

Processing fruit and vegetables

- Primary processing is sorting, trimming, removing damaged produce, peeling, washing and wrapping.
- Fruit and vegetables can be mechanically cored, halved, sliced, chopped, mashed or pitted (stone removed).
- Fruit and vegetables will continue to grow, ripen and decay after harvesting.
- Enzymes cause ripening and decay.
- Freezing prevents decay but is not suitable for soft fruit because ice crystals cause the cell walls to collapse.
- Blanching is plunging fruit or vegetables into boiling water for a few seconds. This slows down decay by destroying enzymes.

> **Exam tip**
>
> Use your knowledge from different parts of the specification to answer questions about the processing of fruit and vegetables.

Figure 12.3 **Methods of preserving fruit and vegetables**

Exam practice answers and quick quizzes at **www.hoddereducation.co.uk/myrevisionnotes**

Primary processing of cereals

- Cereals are the seeds or grains of cultivated grasses.
- Cereals are staple foods, which means they are the main part of the diet for many people.
- The main cereals are wheat, rice, rye, barley, maize and oats.
- Wheat can be grown in a wide range of soils across the world, and is used in breads, pastries, cakes and biscuits.
- Durum wheat is used to make pasta and couscous.
- Rye is a dark colour, has a strong flavour and is used to make crispbread.
- Rice exists in many types: for example, basmati, long grain, pudding rice and brown rice. It requires a hot, wet climate, and is the staple cereal in India and China.
- Maize is used to make cornflour, tortillas, muffins and pancakes.
- Barley is used for making beer and soft drinks, and pearl barley will thicken soups and stews.
- Oats are low in saturated fat; this helps to lower blood cholesterol, which is important in a healthy diet. They are used for oatmeal, porridge oats and jumbo oats.
- The primary processing of cereal grains is to separate the outer layers of the grain from the inner layers; this is done by milling.
- **Milling** is the process of grinding down the cereal grain during primary processing.
- All grains are **screened**, **sorted**, **scrubbed** and **conditioned** (softened with water) before processing.

Structure

- All cereals have a similar structure.
- The three important parts of a grain are the endosperm, bran and germ.
- The purpose of milling is to separate the different parts of the grain.
- The white endosperm needs to be separated from the brown outer bran and the germ.
- Milling breaks down cereal grains between steel rollers.
- Broken grain is sieved and a rough white flour called semolina is collected.
- Semolina and large pieces of grain are fed into reduction rollers.
- Reduction rolling continues until a fine white flour is produced.
- The **extraction rate** is the percentage of flour by weight that is taken from the whole grain to make flour.
- White flour is **fortified** during processing; this means that micronutrients are added to it.
- Regulations state that calcium, iron, vitamin B1 (thiamin) and vitamin B3 (niacin) must be added to white flour.

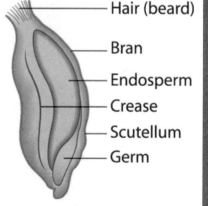

Figure 12.4 **Inside a cereal grain**

- Hair (beard)
- Bran
- Endosperm
- Crease
- Scutellum
- Germ

Table 12.2 **Extraction rate**

Type of flour	Extraction rate	Fibre content
Wholemeal flour	100% whole grain	10% dietary fibre
Brown flour	85% extraction	7% dietary fibre
White flour	72% extraction – all bran and wheat germ are removed	3% dietary fibre

Exam tip

Be specific:
- wholemeal flour is high in dietary fibre
- calcium, iron, vitamin B1 and vitamin B3 are added to white flour.

Types of flour

- The amount of protein in the wheat grain will produce different flours.

Table 12.2 **Protein content**

Type of flour	Protein content	Uses
Plain flour	8% protein	Cakes, biscuits and pastries
Self-raising flour	10% protein	Sponge cakes, scones and puddings
Strong flour	17% protein	Bread, choux pastry, flaky pastry

Secondary processing: flour into bread

REVISED

- The commercial process used in large bakeries to make bread is called the **Chorleywood Bread Process**. Most of the bread in the UK is made this way.

Types of flour for bread making

- Strong flour is used to make bread because it is rich in **gluten**.
- Gluten is a protein that is found naturally in flour. It is developed during kneading in the process of bread making.
- A gluten network is needed to support the carbon dioxide that is produced during the bread-making process.
- Carbon dioxide gives bread a light texture.

> **Typical mistake**
>
> Carbon dioxide gas is produced by fermentation, not oxygen or air.

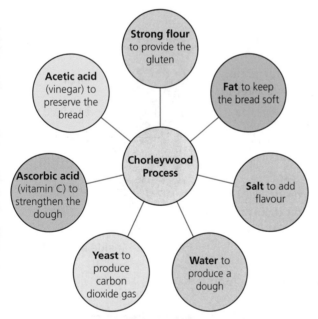

Figure 12.5 Ingredients for bread making using the Chorleywood Process

Figure 12.6 Fermented and baked bread

Bread-making in an industrial bakery:
The Chorleywood Bread Process

● Nine steps are followed during the commercial production of bread (see Figure 12.7).

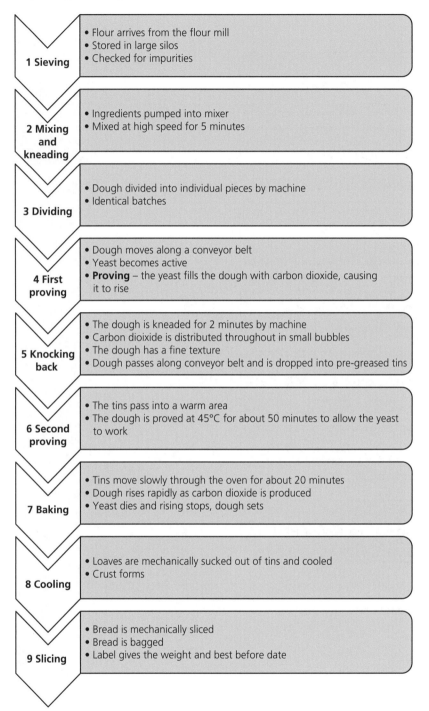

1 Sieving
- Flour arrives from the flour mill
- Stored in large silos
- Checked for impurities

2 Mixing and kneading
- Ingredients pumped into mixer
- Mixed at high speed for 5 minutes

3 Dividing
- Dough divided into individual pieces by machine
- Identical batches

4 First proving
- Dough moves along a conveyor belt
- Yeast becomes active
- **Proving** – the yeast fills the dough with carbon dioxide, causing it to rise

5 Knocking back
- The dough is kneaded for 2 minutes by machine
- Carbon dioixide is distributed throughout in small bubbles
- The dough has a fine texture
- Dough passes along conveyor belt and is dropped into pre-greased tins

6 Second proving
- The tins pass into a warm area
- The dough is proved at 45°C for about 50 minutes to allow the yeast to work

7 Baking
- Tins move slowly through the oven for about 20 minutes
- Dough rises rapidly as carbon dioxide is produced
- Yeast dies and rising stops, dough sets

8 Cooling
- Loaves are mechanically sucked out of tins and cooled
- Crust forms

9 Slicing
- Bread is mechanically sliced
- Bread is bagged
- Label gives the weight and best before date

Figure 12.7 The nine steps of commercial bread production

Secondary processing: flour into pasta

● Pasta is made from a mixture of water and semolina flour made from durum wheat, a yellow-coloured, protein-rich wheat.
● Vegetable juices and herbs can also be added to pasta dough for colour and taste.
● There are seven steps in the commercial production of pasta (see Figure 12.8).

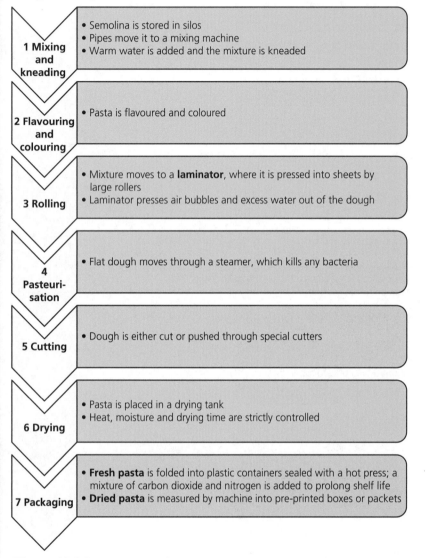

1 Mixing and kneading
● Semolina is stored in silos
● Pipes move it to a mixing machine
● Warm water is added and the mixture is kneaded

2 Flavouring and colouring
● Pasta is flavoured and coloured

3 Rolling
● Mixture moves to a **laminator**, where it is pressed into sheets by large rollers
● Laminator presses air bubbles and excess water out of the dough

4 Pasteurisation
● Flat dough moves through a steamer, which kills any bacteria

5 Cutting
● Dough is either cut or pushed through special cutters

6 Drying
● Pasta is placed in a drying tank
● Heat, moisture and drying time are strictly controlled

7 Packaging
● **Fresh pasta** is folded into plastic containers sealed with a hot press; a mixture of carbon dioxide and nitrogen is added to prolong shelf life
● **Dried pasta** is measured by machine into pre-printed boxes or packets

Figure 12.8 **Seven steps of commercial pasta production**

Primary processing of milk

- Most milk is supplied by Friesian dairy cows.
- Milk is used to make a number of dairy products during secondary processing, including butter, cheese, cream and yoghurt.

The structure of milk

Semi-skimmed milk contains:
- 88 per cent water
- 5 per cent carbohydrate
- 3.5 per cent protein
- 1.5 per cent vitamins and minerals
- 1.7 per cent fat.

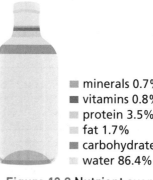

- ■ minerals 0.7%
- ■ vitamins 0.8%
- ■ protein 3.5%
- ■ fat 1.7%
- ■ carbohydrate 4.8%
- ■ water 86.4%

Figure 12.9 Nutrient averages in semi-skimmed milk

Five processing techniques that can be used on milk

1 **Pasteurisation**
   - Pasteurisation is a process in which milk is heated to kill any harmful bacteria that may be present. Most milk is pasteurised.
   - The pasteurisation process used is high temperature short time (HTST):
     ○ the milk is heated to 72°C for a minimum of 15 seconds
     ○ it is then cooled rapidly to below 6°C.

2 **Homogenisation**
   - Homogenisation stops the fat droplets in milk separating out into a layer of cream.
   - It involves forcing milk at high pressure through small holes. This breaks down the fat droplets.
   - The fat droplets spread throughout the milk to prevent separation into a cream layer.

3 **Sterilised milk**
   - Sterilisation is an intense heat treatment and changes the milk a lot. Some people think that sterilised milk tastes 'cooked'.
   - Milk is heated in a steam chamber to a temperature of between 110°C and 130°C for between 10 and 30 minutes.
   - It is cooled and will keep for approximately 6 months without the need for refrigeration.

4 **UHT milk** (ultra-heat treatment)
   - The milk is heated to a temperature of at least 135°C for 1 second.
   - Milk is then packaged into sterile containers.
   - UHT milk has a long shelf life.

5 **Micro-filtered milks**
   - Milk goes through a super-fine filtration system, which removes souring bacteria.
   - The milk is pasteurised.
   - Milk can be stored for 45 days in a refrigerator.

> **Typical mistake**
>
> The heat-sensitive vitamins are destroyed during the heat processing of milk but minerals are not affected.

Secondary processing: milk into cheese

- Most cheese is made from cow's milk but it can also be made from goat and sheep milk.
- There are many different types of cheese, but they are all made using the same process (Figure 12.10).

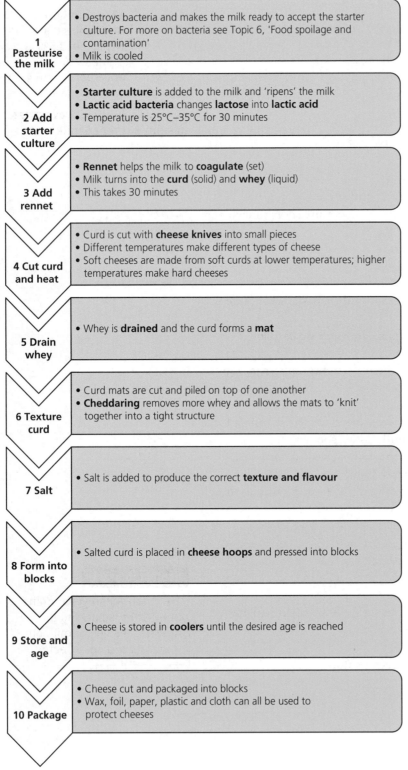

1 Pasteurise the milk
- Destroys bacteria and makes the milk ready to accept the starter culture. For more on bacteria see Topic 6, 'Food spoilage and contamination'
- Milk is cooled

2 Add starter culture
- **Starter culture** is added to the milk and 'ripens' the milk
- **Lactic acid bacteria** changes **lactose** into **lactic acid**
- Temperature is 25°C–35°C for 30 minutes

3 Add rennet
- **Rennet** helps the milk to **coagulate** (set)
- Milk turns into the **curd** (solid) and **whey** (liquid)
- This takes 30 minutes

4 Cut curd and heat
- Curd is cut with **cheese knives** into small pieces
- Different temperatures make different types of cheese
- Soft cheeses are made from soft curds at lower temperatures; higher temperatures make hard cheeses

5 Drain whey
- Whey is **drained** and the curd forms a **mat**

6 Texture curd
- Curd mats are cut and piled on top of one another
- **Cheddaring** removes more whey and allows the mats to 'knit' together into a tight structure

7 Salt
- Salt is added to produce the correct **texture and flavour**

8 Form into blocks
- Salted curd is placed in **cheese hoops** and pressed into blocks

9 Store and age
- Cheese is stored in **coolers** until the desired age is reached

10 Package
- Cheese cut and packaged into blocks
- Wax, foil, paper, plastic and cloth can all be used to protect cheeses

Figure 12.10 **Processing milk into cheese**

Secondary processing: milk into yoghurt

- Yoghurt is available with different textures, fat content and flavours.
- There are different types of yoghurt:
  - live yoghurt contain harmless live bacteria
  - probiotic yoghurt contains 'healthy' probiotic bacteria
  - bio yoghurt has a mild, creamy flavour
  - Greek yoghurt is thick, mild and creamy

The process of commercially making yoghurt

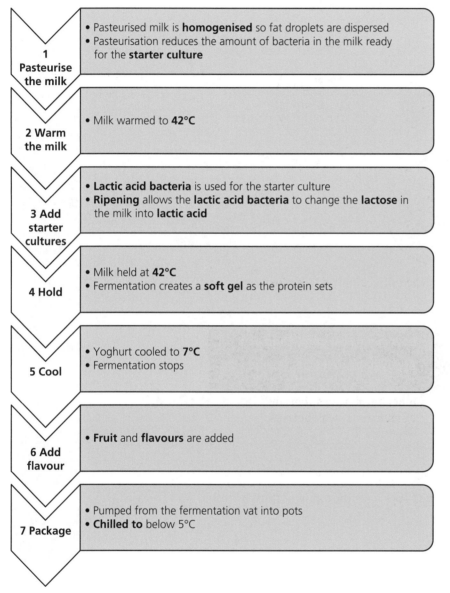

1 Pasteurise the milk
- Pasteurised milk is **homogenised** so fat droplets are dispersed
- Pasteurisation reduces the amount of bacteria in the milk ready for the **starter culture**

2 Warm the milk
- Milk warmed to **42°C**

3 Add starter cultures
- **Lactic acid bacteria** is used for the starter culture
- **Ripening** allows the **lactic acid bacteria** to change the **lactose** in the milk into **lactic acid**

4 Hold
- Milk held at **42°C**
- Fermentation creates a **soft gel** as the protein sets

5 Cool
- Yoghurt cooled to **7°C**
- Fermentation stops

6 Add flavour
- **Fruit** and **flavours** are added

7 Package
- Pumped from the fermentation vat into pots
- **Chilled to** below 5°C

Figure 12.11 Processing milk into yogurt

Secondary food processing: fruit into jam

- **Jam** is made by boiling fruit with sugar until it forms a gel that will set on cooling.
- The **high sugar content** of jam stops micro-organisms from growing.
- **Pectin** is used to set jams.

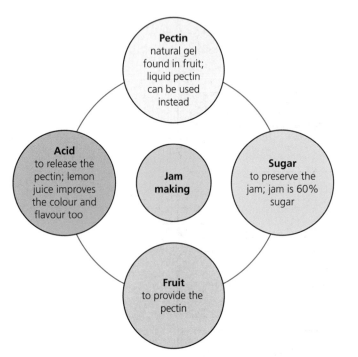

Figure 12.12 The essential ingredients required to make jam

Table 12.3 **Use of pectin in jam**

High in natural pectin and acid No extra pectin and acid is needed	Low in natural pectin and acid Always add extra pectin and acid
gooseberries, citrus fruits, blackcurrants, damsons	blackberries, raspberries, blueberries strawberries, cherries

Problems with jam making

- **Crystallisation** is when sugar crystals appear on the top or within the jam.
- The causes of crystallisation are:
  - too much sugar
  - the sugar not dissolving properly
  - lack of acid.

Equipment for jam making

- Special **preserving pans** are used to make jam. They are usually made of stainless steel, have a long handle, tall sides and a pouring lip.

The process of jam making

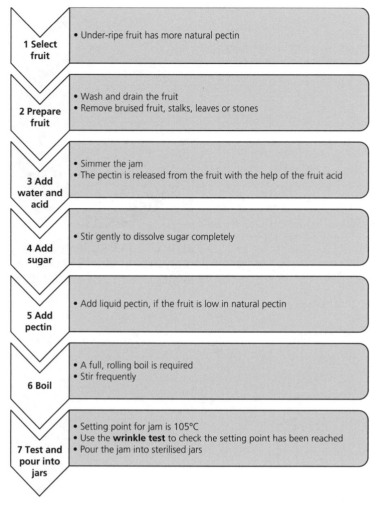

1 Select fruit
- Under-ripe fruit has more natural pectin

2 Prepare fruit
- Wash and drain the fruit
- Remove bruised fruit, stalks, leaves or stones

3 Add water and acid
- Simmer the jam
- The pectin is released from the fruit with the help of the fruit acid

4 Add sugar
- Stir gently to dissolve sugar completely

5 Add pectin
- Add liquid pectin, if the fruit is low in natural pectin

6 Boil
- A full, rolling boil is required
- Stir frequently

7 Test and pour into jars
- Setting point for jam is 105°C
- Use the **wrinkle test** to check the setting point has been reached
- Pour the jam into sterilised jars

Figure 12.13 **The process of jam making**

Food processing and vitamins

REVISED

- Vitamins are lost during every method of food processing.
- Contact with heat, sunlight, water or air will alter the vitamins found in food.
- Vitamin C, vitamin B1 (thiamin), folic acid and vitamin B12 are lost in water.
- Most water-soluble vitamins are destroyed by sunlight and heat.
- Vitamin C is the most sensitive vitamin.
- The food processing techniques that involve heat are **canning**, **pasteurisation** and **sterilisation**.

Table 12.4 **Food processing and vitamins**

Food processing	Foods	Vitamin loss
Canning	Vegetables, fruits	Vitamin C and the vitamin B group are sensitive to heat and are reduced by all these methods
Pasteurisation	Milk, fruit juices	
Sterilisation	Milk	
Drying	Vegetables, fruits	Vitamin C and the vitamin B group are lost when the water is removed
		Vitamin A and E are not water soluble and become concentrated

The effect of heating and drying on the sensory characteristics of milk

REVISED

- The flavour of milk will change depending upon which type of heat treatment it has received.
- The sugar in milk is called **lactose**. It will **caramelise** when heated to a high temperature, causing the milk to taste sweet.
- Sterilised, UHT and evaporated milk will have a sweet, cooked taste due to caramelisation.
- Dried milk tastes slightly sweet or 'cooked' because the milk is heated to a high temperature to remove the water. Caramelisation occurs with heating.
- Homogenisation increases the whiteness of milk because the fat is distributed throughout the milk. The distributed fat will scatter the light better, giving the appearance of whiteness.
- Evaporated milk and condensed milk are a creamy yellow colour.

Figure 12.14 Evaporated milk is a yellowy colour because heating causes caramelisation

Now test yourself

TESTED

1 Describe the different ways that fruit and vegetables can be preserved. [6 marks]
2 Explain what is meant by the extraction rate of cereals when made into flour. [2 marks]
3 List the nutrients used to fortify white flour. [4 marks]

Technological developments associated with better health

Fortified foods and enriched foods

REVISED

Fortified foods are foods that have one or more micronutrients added to them.

- Sometimes foods can be enriched with nutrients. Enrichment means that nutrients are added to improve the nutritional value of a food product.
- White wheat flour is by law fortified in the UK. It is fortified with iron and calcium, and the vitamins B1 (thiamin) and B3 (niacin).
- Micronutrients are found in the bran and wheat germ of the grain that are removed during milling, so they need to be added back in.
- Breakfast cereals are fortified and enriched with vitamins and minerals such as iron, vitamin D, folic acid, riboflavin and niacin. For example, cornflakes have iron added. This makes breakfast cereals more appealing and sometimes replaces the vitamins and minerals lost during processing.
- By law, all butter substitutes, including **fat spreads** and low-fat spreads, must have vitamins A and D added.

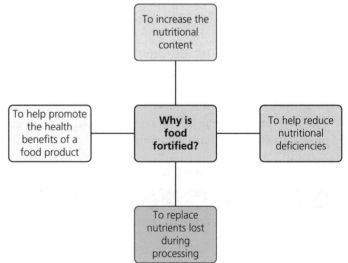

Figure 12.15 Reasons for fortifying food

Foods that lower cholesterol

- Cholesterol-lowering products contain natural extracts from plants, which stop cholesterol from being absorbed into the bloodstream.
- Benecol and Flora pro-activ are **cholesterol-lowering spreads**.

Food additives

- **Food additives** are substances that are added to food during manufacturing.
- All food additives are rigorously tested before they are awarded an E number.
- The European Union has to approve the use of all food additives.

Food additives are added to foods to improve them in different ways. For example:

- to make foods last longer
- to give a wider range of food products
- to improve the flavour or colour of a food product
- to replace nutrients lost during processing
- to maintain the texture and consistency of food products
- to add sweetness without energy or calories.

Table 12.5 Food processing and vitamins

Food additives	Advantages	Disadvantages
Preservatives extend the shelf life of food	Keep food safe by stopping the growth of micro-organisms Reduce food waste Reduce the number of shopping trips	Nitrates have been linked to cancer
Colourings improve the appearance of food (e.g. strawberry yoghurt looks pink)	Replace colour lost in processing Boost natural colour	Some food colours have been linked to hyperactivity in children
Flavourings improve or strengthen the flavour of foods	Restore the flavours after processing Make existing flavours stronger Create products (e.g. cheese and onion crisps)	Monosodium glutamate (E621) may cause symptoms similar to an allergic reaction
Emulsifiers and stabilisers used to mix together ingredients that would not normally mix (e.g. water and oil)	Improve the consistency of food during processing and storage	None are known

Now test yourself

1 List four reasons why food additives are used in food production. [4 marks]
2 Explain the effect of drying on the nutritional value of fruit and vegetables. [3 marks]
3 Which two vitamins must be added to margarine and fat spreads by law? [2 marks]

Revision activity

Draw a web diagram showing the uses, advantages and disadvantages of food additives.

Success in the examination

The written exam

REVISED

In the written exam you will be tested on five topics:
1 Food, nutrition and health
2 Food science
3 Food safety
4 Food choice
5 Food provenance.

The written exam is worth **100 marks** (this is 50 per cent of the total GCSE mark).

Your exam paper will be marked by AQA.

When will the exam be completed?

The written exam will be in May/June of the year in which the qualification is awarded. For most students it will be in Year 11.

How long will I have to complete the exam?

The exam will last for 1 hour and 45 minutes.

What type of questions will appear in the exam paper?

REVISED

The exam will be divided into two sections.
- **Section A** will be **20 multiple-choice questions**. There will be **20 marks** available for this section. For each question there is a choice of four answers with one correct answer for each question. You will need to answer **ALL** the questions in this section.

> **Example**
>
> Example of a Section A multiple-choice question
>
> On the Eatwell Guide, which segment is the largest?
> A Potatoes, bread, rice, pasta and other starchy carbohydrate food
> B Fruit and vegetables
> C Dairy and alternatives
> D Beans, pulses, fish, eggs, meat and other protein

- **Section B** will include five questions. There will be 80 marks available for this section. You will need to answer **ALL** the questions in this section.

> **Example**
>
> Example of a Section B question
>
> Describe a healthy lunch that would supply the micronutrients needed for healthy blood in teenagers. Give reasons for your choice of foods.

The assessment objectives

The assessment objectives (AOs) show what you will be assessed on in the written exam. You should be familiar with the different types of questions so you understand the types of answers required.

	Assessment objectives	Weightings for written exam
AO1	Demonstrate knowledge and understanding of nutrition, food, cooking and preparation	20%
AO2	Apply knowledge and understanding of nutrition, food, cooking and preparation	20%
AO3	Plan, prepare, cook and present dishes, combining appropriate techniques	0%
AO4	Analyse and evaluate different aspects of nutrition, food, cooking and preparation, including food made by themselves and others	10%

The Section A multiple-choice questions are always AO1 questions as they demonstrate your knowledge and understanding.

Section B questions will assess AO1, AO2 and AO4.
- **Example of an AO1 question:** Describe the checks you would make to see if scones were cooked properly.
- **Example of an AO2 question:** Explain the changes that occur when a white sauce thickens during cooking.
- **Example of an AO4 question:** Tooth decay in children is increasing in the UK and has been linked to a diet high in free sugars. Assess the types of foods and lifestyle choices that may contribute to poor dental health in children, and evaluate how healthier choices may improve dental health.

Tips on preparing for the exam

REVISED

- Always ask your teacher if you don't understand something. There is no such thing as a stupid question. Your teacher is there to help you.
- It is never too early to start revising. After each lesson, read your lesson notes and handouts, then make concise revision notes for each topic. The more times you revise a topic, the better you will perform in the exam.
- During the exam, read the instructions on the front cover of the exam paper; these are your instructions for answering the questions.
- Read each question twice to make sure you understand what to do.
- Check how many marks are available for each question. This will tell you how much detail to give in your answers. For example, if there are six marks you may give six different answers, or you may give three answers and give a reason or explanation for each one.
- When you go through the exam paper for the first time, just answer the questions you find easy. If a question seems tricky and you are not sure what to do, leave it and go on to the next question. The second time you go through the paper, answer the questions you are fairly confident with. On the third attempt have a go at answering the trickier questions. Finally, attempt the questions you find very difficult. You should not leave any questions unanswered. A blank space never gets a mark, but sometimes a good guess does!
- **Have a positive attitude. Do not allow self-doubt to affect your preparation or success.**

Approaching the paper

- The day before your exam, check that you know the time and place of your exam. Make sure you have the correct items in your pencil case. Have an early night so you are alert.
- Give yourself time to answer all the questions. You should aim to spend about 20 minutes on Section A of the paper and about 1 hour 25 minutes on Section B. Within this time, you should check your answers by re-reading the questions and making sure you have answered them correctly.
- Command or trigger words will give you a clue to the depth of the response required. For example:
  - ○ **State** would mean you just need to give the bare facts.
  - ○ **Explain** means you need to give reasons for the facts to make them plain and clear.
- Highlight these command words so you don't go off the point.

State	Give only the bare facts, expressed clearly and fully.
Select	Carefully choose as being the best or most suitable.
Identify	Establish or indicate what someone or something is.
Suggest	Make a recommendation or suggestion.
Describe	Write out the main features. Write a picture in words.
Outline	Write out the main points or a general plan, but omit minor details.
Explain	Set out facts and the reasons for them, make them known in detail and make them plain and clear.
Consider	Think about in order to understand or decide.
Justify	Show adequate grounds for decisions or conclusions. Prove to be right. Give a good reason.
Compare	Point out the differences and similarities between the given items.
Contrast	Point out the differences between two or more given items.
Discuss	Write from more than one viewpoint, supporting and casting doubt. It is not always necessary to come to a conclusion.
Assess	Give your judgement of something. Put a value on it. Judge the worth of something.
Evaluate	Judge the worth of something, sum up the good and bad parts and decide how improvements may be made.
Draw conclusions from	Explain what you learnt.

Sample examination questions

Section A: Multiple-choice questions

Example

Too much fat in our diet can lead to:
(a) Tooth decay and obesity   (c) Obesity and heart disease
(b) Anaemia and scurvy   (d) Diabetes and rickets

Commentary

Each multiple choice questions will be worth one mark.
You will be given a question with four potential answers and will need to select one correct answer only.
If you do not select the correct answer you will not be awarded a mark.

Section B questions

Sample question 1: AO1 assessed question

Example

Explain how the micro-organisms – yeasts, moulds and bacteria – can spoil food.   [6 marks]

Response 1 High-level response – total 6/6 marks

The micro-organisms may cause food spoilage in a wide range of foods. These begin the natural decay process of food.

Yeasts are single-celled fungi that reproduce by budding. Yeasts can grow in acidic, sweet food such as orange juice and may cause this to ferment. Yeasts grow best in warm conditions around 25–29°C.

Moulds are tiny fungi that produce thread-like filaments that help the mould to spread around the food. Moulds grow in warm (between 20 and 30°C) and moist conditions. They spoil food such as bread and cheese, and soft fruits such as strawberries.

Bacteria that cause food to decay are called food spoilage bacteria. These can cause food to smell and lose its texture and flavour. When bacteria break down foods they produce acids and other waste products. Bacteria prefer body temperature to multiply, around 37°C.

Response 2 Low-level response – total 2/6 marks

Yeasts can grow on foods like oranges and spoil their appearance and flavour.

Moulds grow easily on bread and spoil its taste and appearance. Some moulds grow in the fridge.

Bacteria can cause food to decay and break down.

Commentary

When seeing an 'explain question' the student should think about setting out the facts and the reasons for them, to make them clear. There should be sufficient detail for top-band marks from each of the three micro-organisms, showing a comprehensive knowledge and understanding of how these may spoil food.

Response 1: 6/6 marks

Sample 1 is a high-level response.

The student shows correct and comprehensive knowledge of yeasts, moulds and bacteria. The conditions they require are explained, as well as the spoilage that may occur. There is appropriate and accurate use of specialist terminology. There are accurate, concise and detailed explanations of all three micro-organisms for this 6-mark question.

Response 2: 2/6 marks

Sample 2 is a low-level response.

The student shows some correct knowledge of how the micro-organisms may spoil foods and gives some basic explanations. Yeasts and moulds are provided, with an example of a food that may be affected, but the bacteria sentence, although correct, is rather vague and has no example of a food that may be affected by bacterial spoilage.

Mark scheme

This question is assessed against AO1.

Students will recall and relate knowledge and understanding of how yeasts, moulds and bacteria may spoil food.

Marks awarded for each section as follows.

Band	Descriptor
High-level response 5–6 marks	Response identifies the correct and comprehensive knowledge and understanding of how yeasts, moulds and bacteria may spoil food. Work will show accuracy and use a range of specialist terminology correctly.
Middle-level response 3–4 marks	Response identifies the correct knowledge with a good explanation of how yeasts, moulds and bacteria may spoil food. Work will include the occasional inaccuracy but will use most specialist terminology correctly.
Low-level response 1–2 marks	Response identifies some correct knowledge with a basic explanation of how yeasts and/or moulds and/or bacteria may spoil food. Work will include the occasional inaccuracy, but will use some specialist terminology correctly.
0 marks	Incorrect knowledge or no specific examples of how yeasts, moulds and bacteria may spoil food.

Indicative content

Yeasts

- Yeasts are single-celled fungi that reproduce by 'budding'.
- Can grow in acidic, sweet foods (e.g. orange juice can ferment if it is not stored correctly).
- Yeast may ferment honey that is not pasteurised.
- Budding means the yeast cell grows a bud that becomes bigger until it eventually breaks off and becomes a new yeast cell.
- Yeasts can grow with (aerobic) or without (anaerobic) oxygen.
- Yeasts prefer moist, acidic foods.
- Yeasts can grow in high concentrations of sugar and salt.
- Yeasts grow best in warm conditions (around 25–29°C).
- Yeasts can grow at fridge temperatures (0 to below 5°C).
- Yeasts are destroyed at temperatures above 55°C.

Moulds

- Moulds are tiny fungi that produce thread-like filaments that help the mould to spread around the food.
- Moulds grow in warm and moist conditions.
- Mould grows easily on bread, cheese and soft fruits.
- Moulds can grow in foods with high sugar and salt concentrations.
- Moulds grow best between 20°C and 30°C.
- Moulds can grow slowly in the fridge at 0 to below 5°C.
- Mould growth may be speeded up by high humidity and fluctuating temperatures.

- Moulds can grow on fairly dry food, such as hard cheese (e.g. Cheddar cheese).
- Moulds often spoil food such as bread and other bakery products.

Bacteria
- Some types of bacteria cause food to decay; these are called food spoilage bacteria which cause food to smell and lose their texture and flavour.
- When bacteria break down the food, acids and other waste products are created in the process.
- These bacteria can sometimes make the food product unsafe to eat.
- Pathogenic bacteria can cause food poisoning.
- Bacteria grow best with warmth (around body temperature 37°C), they prefer moist conditions on neutral foods containing protein.
- Bacteria do not like high concentrations of salt, sugar or acid.

Sample question 2: AO2 assessed question

Example

Describe the characteristics of an international cuisine of your choice.
[9 marks]

Response 1 High-level response – total 9/9 marks

The Italians enjoy social occasions where food can be shared; food is presented simply. The Italians love to have three or more courses for their meals over a long period of time.

Traditionally breakfast is usually light, consisting of coffee and some bread rolls. This is then followed by two main meals – one is served in the middle of the day, the other in the evening, usually quite late.

Northern Italy is cooler and more mountainous than the south so rice is grown and the land is mostly used to rear animals for cured meats. Southern Italy is hotter, so more crops are grown such as lemons and olives.

Pigs are reared for meat, which due to the hot climate is cured and made into products such as salami and pancetta. Fish such as shellfish and squid are used in traditional dishes such as risotto, which may be sprinkled with Parmesan cheese.

Traditional Italian dishes include gnocchi, which is a potato dough made into dumplings and usually served with a tomato sauce. Pizza is a bread dough base covered with tomatoes and mozzarella cheese, then baked. Pasta dough is made into a variety of shapes, often using a pasta machine, making dishes such as lasagne, cannelloni and ravioli. Pasta dishes are boiled or baked.

Traditional Italian desserts are tiramisu, amaretto, ice cream (gelato) and pannacotta, a cooked cream dessert set with gelatine.

The Italians enjoy coffee such as espresso and cappuccino.

Response 2 Low-level response – total 2/9 marks

Italy has lots of different foods as it is a long country with lots of sea around it. Lots of fish are caught and this is used in many recipes. In Italy they like to eat pizza and pasta with different sauces. Italians like ice cream for pudding because some parts of Italy are very hot, especially in the summer. In Italy they eat their dinner really late compared to England.

Commentary

When seeing a 'discuss question' the student should write from more than one viewpoint. There should be sufficient detail for top-band marks to show an accurate knowledge of an international cuisine, with relevant examples that show clear understanding; along, with specialist terminology used accurately.

Response 1: 9/9 marks

Sample 1 is a high-level response.

This high-band response shows accurate application of knowledge and relevant examples of an international country's cuisine, and provides detailed examples of the characteristics of that country's cuisine. The answer also shows a range of specialist terminology used correctly.

Response 2: 2/9 marks

Sample 2 is a low-level response.

This answer scores in the low mark band as it shows a limited application of knowledge with few examples. It is a basic response but has some qualified answers and gives examples of some international foods.

Mark scheme

This question is assessed against AO2.

The characteristics are:
- distinctive features of the cuisine
- traditional foods grown/reared/caught
- traditional dishes
- main cooking methods
- specialist equipment used for cooking
- eating patterns.
- presentation styles.

Marks awarded for each section as follows.

Band	Descriptor
High-level response 7–9 marks	Response shows accurate application of knowledge of an international cuisine with relevant examples, which show a clarity of understanding.
	Responses include detailed factual explanation with qualified answers.
	Work is accurate and shows a range of specialist terminology used correctly.
Middle-level response 4–6 marks	Response shows some application of knowledge of an international cuisine with appropriate examples given.
	A grasp of most aspects is shown, but some areas lack clarity.
	Responses include factual responses with some explanation and qualified answers.
	Work will include occasional inaccuracy but will use most specialist terminology correctly.
Low-level response 1–3 marks	Response shows limited application of knowledge of an international cuisine.
	Few examples will be given, showing a grasp of some aspects but areas may lack clarity.
	Responses will include basic responses and few qualified answers.
	Work will include the occasional inaccuracy but will use some specialist terminology correctly.
0 marks	No answer worthy of credit.

Sample question 3: AO4 assessed question

Example

Discuss the use of food additives in processed foods.   [12 marks]

Response 1 High-level response – total 12/12 marks

Food additives are very important in the production of high-quality and safe foods. Preservatives are vital because they keep food safe to eat for longer and help us to reduce food waste. For example, nitrites are used in bacon and ham to stop bacteria from growing.

Food additives can also improve the appearance and flavour of food. Food colourings and flavourings are used to make food look and taste better. For example, strawberry yoghurt may have colour and flavour added to make it taste and look like real strawberries have been used. These colours and flavours can be natural or artificial.

Emulsifiers prevent foods from separating and give food texture. For example, the emulsifier lecithin is found in mayonnaise. Stabilisers can also be used in mayonnaise to prevent the mayonnaise emulsion from separating. The use of emulsifiers and stabilisers is important because the consumer has more choice of food products to eat.

There are health concerns about the use of some additives that are allowed to be added to food. It has been suggested that a food colouring called Sunset Yellow found in squashes may cause hyperactivity and behavioural problems in children. Other additives have been linked to the increase in asthma in children that has taken place during recent years.

Response 2 Low-level response – total 4/12 marks

Food additives are used to improve food. There are many different types of food additives. They can improve the colour and taste of food. Colourings make sure that strawberry yoghurt looks pink so that people will want to eat it. Flavourings also give crisps a strong flavour like cheese and onion. Additives can be used to make the food look and taste better. Food additives will also stop food from going off. Preservatives are used in tinned foods and help them keep for longer. Some additives stop things from breaking up and work in sauces.

Food additives are made in factories and are man-made chemicals. They are used in large amounts by food companies. They are harmful to health.

Colourings cause hyperactivity in children. E numbers should be tested.

Commentary

When seeing a 'discuss question' the candidate should think about giving an account that addresses a range of ideas and arguments. They should attempt to write from more than one viewpoint, supporting and casting doubt. The response should include an element of evaluation but it is not always necessary to come to a conclusion.

Response 1: 12/12 marks

Sample 1 is a high-level response.

In the separate paragraphs the candidate clearly explains the importance of several different types of food additive. There is appropriate and accurate use of specialist terminology throughout, and accuracy of spelling, punctuation and grammar is very good. There are accurate and detailed factual explanations of five food additives:

- preservatives are mentioned and supported by relevant examples – 'nitrites are used in bacon, ham to stop bacteria from growing'
- food colourings and flavourings are stated to make food more attractive, with an example of strawberry yoghurt, which contains both of these additives
- an understanding of emulsifiers and stabilisers is included, with a supporting example of lecithin in mayonnaise.

There is a paragraph devoted to the shortcomings of food additives with named examples. Three health concerns are mentioned: hyperactivity

in children, behavioural problems and asthma. But the evaluative suggestion is that their value is important. There is a good balance between analysis and evaluation.

Response 2: 4/12 marks

Sample 2 is a low-level response.

There is limited factual explanation, showing basic knowledge of the indicative content. The candidate has mentioned three food additives but is unable to access the middle band because there is limited reference to their use. Only colourings and flavours have any analysis and there is limited specialist terminology:

- colourings make sure that strawberry yoghurt looks pink so that people will want to eat it
- flavourings also give crisps a strong flavour like cheese and onion
- food additives will also stop food from going off
- some additives work in sauces.

There are inaccuracies in the information presented. Preservatives are not generally used to preserve tinned foods. There are generalisations in the expression 'Colourings cause hyperactivity in children.' But the accurate reference to hyperactivity would be credited. There may be an imbalance between analysis and evaluation. Only a limited conclusion, with little attempt to link the use of food additives to potential health risks, is provided.

Mark scheme

This question is assessed against AO4.

Marks awarded for each section as follows.

Band	Descriptor
High-level response 9–12 marks	Response will include accurate and detailed factual explanations, showing thorough knowledge of food additives linked closely to the indicative content. Appropriate and accurate use of specialist terminology. There will be a good balance between analysis and evaluation. Analysis is excellent, and accurately identifies and describes the importance of three or more food additives. Makes reference to the use of food additives. Evaluation will make sound judgements, linking the use of food additives to at least two potential health risks.
Middle-level response 5–8 marks	Response will be mainly accurate, with some factual explanations showing good knowledge of nutritional issues linked to the indicative content. Appropriate and good use of specialist terminology. There will be a reasonable balance between analysis and evaluation. Analysis is good, and identifies and describes the importance of at least two food additives. Makes some reference to the use of food additives. Evaluation will make some judgements, linking the use of food additives to at least one potential health risk.
Low-level response 1–4 marks	Response will include limited factual explanations showing basic knowledge of nutritional issues linked to the indicative content. There may be a limited attempt to use specialist terminology. There may be an imbalance between analysis and evaluation, where one aspect may be omitted or stronger. Analysis is limited and identifies one or two food additives, with minimal or no description of their importance. Little or no reference to the use of food additives. Evaluation will make limited judgements, with little attempt to link the use of food additives to potential health risks.
0 marks	No answer worthy of credit.

Indicative content

Food additives

- Candidates may raise concern about the overuse of food additives and the potential dangers of their use. This must be qualified. There is very little scientific evidence that food additives used in the EU pose any threat to health. Food additives are strictly tested for safety.
- Food additives allow new products/varieties to be developed and exist (e.g. ice creams, fat spreads).

Colours

- To make food look attractive, meet consumer expectations and restore colour, which may have been lost during processing of some food, additives are used to restore the original colour.
- Colour additives can also be used to make the existing food colour brighter (e.g. to enhance the yellowness of custard).

Preservatives

- Used to maintain freshness and stop the growth of micro-organisms. Food can be transported for greater distances and stored longer (e.g. sulphur dioxide).
- Are used to help keep food safe for longer.

Flavourings

- To improve a specific characteristic of a food product (e.g. strawberry yoghurt with a strawberry flavour).
- Used widely in savoury foods to make the existing flavour in the food stronger (e.g. monosodium glutamate).

Emulsifiers, stabilisers

- Emulsifiers and stabilisers allow fat and water to be mixed together to create low-fat spreads. They also give food products a smooth, creamy flavour.
- Emulsifiers help mix together ingredients like oil and water that would normally separate; stabilisers prevent them from separating again.
- Gelling agents are used to give foods a gel-like consistency, while thickeners increase the thickness of foods (e.g. pectin in jams).

Nutrient enrichment

- Vitamins and minerals are added to products to replace the nutrients lost during processing (e.g. breakfast cereal).
- Vitamins are added to give health benefits to certain groups who may suffer a deficiency.

Exam practice answers and quick quizzes at **www.hoddereducation.co.uk/myrevisionnotes**